HOW TO REPLACE YOUR FIRE CHIEF WITH AI

A Story-Driven Guide to Leadership and AI for the Fire Service

HOW TO REPLACE YOUR FIRE CHIEF WITH AI

A Story-Driven Guide to Leadership and AI for the Fire Service

Alex Zielinski

Paperback ISBN: 979-8-9966136-0-1
Digital ISBN: 979-8-9966136-1-8

A Note Before You Begin

The characters in this book — Chief Dana Mercer, Chief Marcus Webb, Battalion Chief Priya Nair, ChiefBot 9000, and everyone they encounter — are fictional. They are composites designed to illustrate the dynamics of people in specific situations rather than document real people. The circumstances and events they inhabit are illustrative, not historical or biographical.

The AI capabilities described in this book are real. The fire service data and research on leadership that underpin the arguments are real. The organizational dynamics — the cultures, the politics, the crises, the failures of presence — are real in the sense that any experienced fire service leader will recognize them, because they are drawn from the patterns of a profession rather than the particulars of any single department.

The story is the delivery mechanism. The argument is what the story carries.

Not Sure What AI Actually Is? Start Here.

If you already work with AI tools regularly, feel free to skip ahead. This page is for everyone else — and there is no shame in being here.

Artificial intelligence is not a robot. It isn't sentient. It doesn't think, feel, or want anything. AI is pattern recognition at enormous scale — software processing vast amounts of data to identify what comes next, what fits, what matches. In a way, it's a probability and prediction machine. When you ask an AI tool to draft a report, it is not composing the way a human does. It's predicting, with remarkable accuracy, what a good report looks like based on millions of examples it processed.

That's both more impressive and less frightening than most people assume.

There are a few terms worth knowing before you read this book. None of them require a technical background. They just need a definition.

Artificial Intelligence (AI): A broad term for software systems designed to perform tasks that typically require human-like reasoning — recognizing patterns, making predictions, generating language, or solving problems.

Machine Learning: A type of AI that improves through experience. The more data it processes, the better it gets at its task without being explicitly reprogrammed.

Large Language Model (LLM): The technology behind tools like ChatGPT, Claude, and others. Trained on enormous amounts of text, these systems generate remarkably coherent written responses to prompts. This is the category of AI most relevant to the work of a fire chief.

Automation vs. Augmentation: Automation replaces a task entirely. Augmentation enhances a human doing the task. This distinction matters enormously throughout this book — and the two are far more often confused than well-defined and articulated.

Prompt: The input you give an AI tool. A vague prompt produces a vague result. A specific, well-crafted prompt produces something genuinely useful. Learning to write good prompts is a learnable skill, and it's not complicated.

That's enough to get started. Everything else will make sense in the context of the people and story ahead.

Introduction

The Memo

Scene

The memo arrives at 6:47 a.m. on a Tuesday.

ChiefBot 9000 — officially designated the Municipal Leadership Optimization System, Version 4.2, though nobody calls it that — receives it, processes it, and generates a response in 0.3 seconds. The response is professional, appropriately enthusiastic, and contains no grammatical errors. ChiefBot files a copy in the administrative record, flags three action items for follow-up, and adds the pilot program launch date to the department calendar. It does not pause to consider the memo's implications. It doesn't notice the weight of what it has just been asked to do. It logs the request and moves on.

By 7:15 a.m., ChiefBot has drafted a 90-day implementation plan, identified four budget line items that could be reallocated to support the transition, and generated a stakeholder communication template for the city council briefing. The plan is thorough. The budget analysis is accurate. The communication template is, objectively, better than anything the previous chief's administrative assistant produced in fourteen years.

Chief Dana Mercer reads the memo at 7:22 a.m., standing at the kitchen counter with her second cup of coffee going cold in her hand.

She reads it once. Sets it down. Picks it up again.

The subject line says: Pilot Program — Municipal Leadership Optimization Initiative

The body says: Efficiency. Optimization. Technology-forward leadership.

She has been a fire chief for eleven years. She built this department's training program from a three-ring binder and institutional memory. She's sat with families at 3 a.m. She's testified before city council with her hands shaking under the table and her voice steady above it. She's navigated a union grievance that nearly split the department in half. She's done every hard thing this job has ever asked of her.

She has never once described any of it as optimization.

She sets the memo face-down on the counter. Picks up her coffee. It's cold. She drinks it anyway.

She stands there for a moment, looking at the memo face-down on the counter. She's thinking about something she cannot quite name yet — a shape in the air that's not quite fear and not quite anger and is something more like the specific disorientation of someone who has just been told that the work of their life is a problem to be solved. She's been a fire chief for eleven years. She's solved problems. She, until now, has never thought of herself as one.

Across town, the same memo lands in Chief Marcus Webb's inbox at 7:31 a.m., forwarded by a colleague with a subject line that says only: have you seen this.

Marcus has seen versions of this memo before. Different cities. Different language. Same question underneath: prove your value or make room for something cheaper.

He answered that question before, in three departments across twenty-two years. Not with a memo of his own. With the quality of his work and the depth of his relationships and the specific institutional knowledge that made him, in each building, the person the organization could not afford to lose. He has learned, through iteration, that the answer to prove your value is not an argument. It is a demonstration.

He sets his phone down. Picks up his coffee — still hot, he made it two minutes ago — and opens his laptop.

He types three words into a search bar.

What can't AI do?

He reads for twenty minutes. He takes three notes. Then he closes the laptop, finishes his coffee, and starts getting ready for work.

The answer, it turns out, is the reason this book exists.

Story

This book is not about technology.

It's about leadership — specifically, about what fire service leadership actually requires, what it costs when it's done poorly, and what it looks like when it's done well in an era of rapid, uncomfortable change. The title is a provocation, not a forecast. Artificial intelligence will not replace your fire chief. But it can, already, do a striking number of things your fire chief does every day — and that fact carries implications worth taking seriously.

The story you are about to read follows Dana Mercer through the year that ChiefBot 9000 arrives in her department. It's not a comfortable year. It is the kind of year that asks you to look at yourself honestly, which is the kind of year most people spend considerable energy avoiding.

Dana is not a bad chief. She's a good chief who has stopped growing. Her competence is real. Her institutional knowledge is genuine. Her love for the department is not in question. What is in question is whether she has the self-awareness and the will to lead differently in a world that's changing faster than she's been willing to acknowledge.

That question is this book.

You will also follow Marcus Webb — a chief who has worked four departments across three states in twenty-two years, who has been the new person enough times that 'new' feels familiar, who does not experience change as a threat because he has never had the luxury of expecting the environment to stay the same. Marcus isn't the hero of this book. He is the proof that the answer exists. He is what becomes possible when a leader has exercised, repeatedly and under pressure, the muscle that Dana has allowed to atrophy.

Neither Dana nor Marcus is you, exactly. But you might find yourself in both of them — in Dana's defensiveness, in Marcus's adaptability, in the gap between where you are and where you know you should be. The question the book keeps asking is not which of them you resemble. It's what you are going to do about it.

And then there's ChiefBot 9000.

ChiefBot is the system the city manager installed to optimize fire department leadership. It's not malicious. It's not brilliant. It is *relentless*. It's unsettlingly competent at everything it can measure — and completely without awareness of everything it cannot. It doesn't hesitate. It does not notice. It processes, generates, flags, and moves on, indifferent to the weight of what it has just done.

It will, over the course of this book, do many things that Dana expected to be hers. Each time, she'll have to ask herself the question the city manager's memo put into the air but did not have the honesty to say out loud:

What, specifically, do you do that a machine cannot?

The answer isn't obvious. It's not flattering in all its parts. And it's the most important question a fire service leader can ask in the decade ahead.

This book is organized in three parts. Part One follows Dana through four chapters in which ChiefBot does her job — administrative work, data analysis, training assessment, strategic planning — with precise unsettling competence, because it's genuine. These chapters are not an argument for AI. They are *an honest account* of what it can already do, because a reader who doesn't understand that cannot make sense of what follows.

Part Two follows Dana through four chapters in which ChiefBot fails — not because it's poorly designed, but because the situations require something it cannot have: presence, relationship, political judgment, ethical discernment. These chapters are the book's core argument, and they're told as specifically and honestly as the story allows.

Part Three is about what comes next. How you build the habits and the workflows that let you lead effectively in this environment. How you stay relevant when the tools keep improving. How you find the people who are already living in the future and learn from them rather than resenting them.

There is a Bridge between Part Two and Part Three. It's the most important page in the book. Read it slowly.

If you picked up this book feeling defensive, that reaction is worth examining. The fire service has a long and honorable tradition of doing things the right way, and a complicated relationship with doing things a new way. Those two things are not the same, though they're often confused for each other. Institutional knowledge is real and valuable. So is institutional inertia. The challenge — and the argument of this book — is learning to tell the difference.

If you're already convinced that AI is the future and anyone resisting it is a dinosaur, slow down. The picture is more nuanced than that, and the nuance matters. Experience, relationships, judgment, and presence are not soft skills. They are the core of what a fire chief is for, and no algorithm replaces them. The chiefs who understand this — who use technology precisely because they know what technology cannot do — are the ones who will lead the strongest departments of the next decade.

This book is for both of you. And for everyone in between.

The story begins on a Tuesday morning, with a memo nobody asked for and a question nobody is quite ready to answer.

By the end, Dana Mercer will have an answer.

Whether she acts on it is up to her. Whether you act on yours is up to you.

How to Read This Book

Each chapter opens with a scene — a moment in Dana's year with ChiefBot. The scenes are followed by the story section, which explains what that moment reveals: about AI, about leadership, about the gap between the chief Dana has been and the chief she needs to become.

At the end of each chapter, a section called **For the Reader** steps briefly outside the story to make the practical argument explicit. These sections are short by design — the story has already done the teaching. But they are designed to be used, not skimmed. Each one contains questions worth sitting with and a challenge worth accepting. If you find yourself reading past them quickly, go back. The answers you give in those sections are the point of the exercise.

You don't need to agree with everything in this book. You need to engage with it honestly.

That's the only requirement.

Let's begin.

PART ONE

What AI Can Do (Frighteningly Well)

Chapter 1

The Paper Chief

Scene

It is 11:47 p.m. on a Thursday.

The station is quiet in the way fire stations get quiet at this hour — not silent, never silent, but held in the particular suspended readiness that experienced firefighters learn to inhabit without thinking about it. Somewhere in the building a television murmurs. The apparatus bay breathes the faint smell of diesel and metal polish. Two engines and a ladder truck sit in the dark, waiting.

ChiefBot 9000 isn't waiting.

It does not wait, and it does not rest. ChiefBot does not experience the particular administrative dread that descends on a fire chief every Sunday evening when the week's undone work surfaces in memory like a slow-rising tide. Since the department's administrative server came back online four hours and thirteen minutes ago, ChiefBot has been running.

The ISO documentation package: complete. Forty-seven pages, fully cross-referenced, supporting data pulled directly from the records management system. Three policy documents flagged as unreviewed for more than twenty-four months, updated drafts generated, queued for chief officer review. Monthly training compliance report generated, color-coded by company, automatic notifications drafted for the two members whose recertifications lapse within sixty days. Next quarter's shift schedule built from seven months of leave request history, holiday rotation patterns, and minimum staffing requirements — a conflict in Week 11 identified and resolved before it became anyone's problem. Apparatus maintenance log updated. Overtime report reconciled. Four routine administrative emails drafted, ready for the chief's review and signature.

Total time elapsed: four hours, twelve minutes.

ChiefBot flags the completed items, generates a summary report, and queues it for Chief Mercer's 0630 review. Then it begins the next task, without pausing, without noticing what it's just done.

Dana arrives at 0631.

She has been in the fire service for twenty-seven years. She spent sixteen of them working her way up — probationary firefighter, engineer, lieutenant, battalion chief — before she ever sat in the chief's chair. She has been chief for eleven.

She sets her coffee on the desk — the same desk, the same mug, eleven years of the same routine — and opens her laptop. The summary report is at the top of her queue. She reads the subject line. Reads it again.

She opens the report.

She reads for six minutes without moving. When she reaches the end she does not feel what she expected to feel — she is not sure, precisely, what she expected to feel. Gratitude would be the reasonable response. The work is done. It is done correctly. It is formatted correctly and filed correctly and she does not have to do any of it. She should feel gratitude.

She feels something else.

She closes the laptop. Picks up her coffee. Walks out to the apparatus bay.

She stands in the half-dark between Engine 1 and the ladder truck, the concrete cold through the soles of her boots, and looks at the trucks. She has been looking at trucks her whole career. She started as a probationary firefighter in a department three counties over, riding the back of an engine that smelled exactly like this one, and the sight of apparatus — the specific combination of mass and readiness and purpose, the compressed potential of a machine built for one thing and very good at it — has grounded her in every difficult moment she has navigated in twenty-seven years of this work.

It is not grounding her right now.

She thinks about the ISO package. Forty-seven pages. She had been planning to spend the better part of next week on it — pulling records, cross-referencing criteria, formatting the submission. She had already rearranged two meetings to protect the time. She had been carrying the task in her head for three weeks, the way you carry a task that is large and necessary and not yet started, that unfinished weight that surfaces at inconvenient moments and reminds you it is still there.

It is not there anymore. It is done.

She takes a sip of coffee. It has gone cold.

The work is done. All of it. And she does not know what to do with that.

Story

If you have been a fire chief for more than six months, you know the feeling Dana is standing in.

Not anger. Not gratitude. Something more complicated than either — a quiet vertigo that comes from watching something do in four hours what consumed the better part of your last three weeks. The work is done. It is done correctly. It is, if you are honest with yourself, done better than you would have done it — because you would have done it in pieces, between interruptions, at the end of long days when your best thinking was already spent on everything else.

Dana has been a fire chief for eleven years. In eleven years she has never arrived at her desk to find her work done.

This is going to require some adjustment.

The administrative burden of fire service leadership is one of the profession's most poorly kept secrets. Chiefs talk about it at conferences the way people talk about a chronic condition they have learned to manage — with dark humor, a certain resignation, and the unspoken assumption that it simply comes with the territory. ISO documentation. NERIS reporting. Policy review cycles. Accreditation prep. Budget narratives. Compliance tracking. Training records. Personnel files. Apparatus maintenance logs. Mutual aid agreements. Grant reporting. The list is not static. It grows, year over year, as regulatory requirements expand, accreditation standards evolve, and the documentation demands placed on fire departments continue to accumulate.

Research on managers across industries consistently finds that administrative tasks consume roughly one full day per week. Fire chiefs carry a version of this burden that most comparable leadership roles don't — layered with ISO documentation cycles, NERIS reporting, accreditation prep, training compliance tracking, and a grant landscape that requires sustained attention to navigate. Ask any experienced chief to estimate what percentage of their professional week goes to paperwork. Whatever number they give you will surprise them when they say it out loud.

This is not a character flaw. It is a structural problem. And it is one that AI is already solving, right now, for chiefs willing to use it.

What ChiefBot Actually Did

ISO documentation is one of the most labor-intensive recurring administrative tasks in the fire service. The Insurance Services Office evaluates departments across hundreds of data points — water supply, equipment, training records, communications systems, company distribution — and the documentation required to support a competitive ISO rating can consume weeks of staff time. AI tools can now pull structured data directly from records management systems, cross-reference it against ISO criteria, identify gaps, and generate formatted documentation packages. Departments are doing this today.

Policy review and updating is another area where AI performs with uncomfortable competence. Most departments have policy manuals running to hundreds of pages, with review cycles that are supposed to happen annually and actually happen whenever someone has time — which is to say, irregularly. AI tools can be configured to flag policies by review date, draft updated language based on current best practices and regulatory changes, and queue documents for human review and approval. The chief still makes the final call. The AI does the legwork.

Scheduling, compliance tracking, and routine administrative correspondence follow the same pattern: AI handles the volume, the pattern recognition, the continuous monitoring. It does not get tired. It does not forget that Engine 3's driver has a standing medical appointment on the third Tuesday of every month. It does not let four routine emails sit in an inbox for a week because other things kept coming up.

None of this requires advanced technical knowledge. None of it requires a significant budget. Much of it is available through tools that cost less per month than a department's coffee budget.

Dana reads the summary report a second time at her desk, coffee reheated, an hour after she came in from the apparatus bay. She is looking for errors. She does not find any.

She finds herself reading the schedule section three times. She built the last four quarters' schedules by hand — a process she had refined over years into something that felt almost like a skill, an expertise, a thing she was specifically good at. The color-coded version she maintained in a spreadsheet that she had never shared with anyone because explaining it would have taken longer than just doing it herself. The process of building a schedule she understood as a form of knowing the department: who could work with whom, who needed accommodation, which crews had unspoken tensions that manifested in overtime requests, which battalion chief would fight her on certain assignments and which would not.

ChiefBot's schedule is correct. It has identified a conflict in Week 11 that she would eventually have caught and resolved. It has no opinion about which crews have unspoken tensions. It has built a technically flawless schedule and it knows nothing about the department it has scheduled.

She closes the laptop again. Sits with that for a moment.

Both things are true. That is the part she is going to have to get used to.

The Scheduling Officer and the Spreadsheet

Marcus Webb is a fire chief who has worked in four departments across three states in twenty-two years. The breadth of that experience has given him something that Dana, eleven years in one building, does not yet have: a practiced relationship with being the new person.

At his third department, Marcus inherited a deputy chief who had been building the shift schedule by hand for nine years. The deputy's schedule lived in a color-coded spreadsheet of remarkable complexity — a document the deputy had never shared electronically, printed and brought to scheduling conversations in a manila folder, and regarded with the particular quiet pride of someone who has built something functional from nothing and knows it.

Marcus looked at the spreadsheet for fifteen minutes during his first week. He did not say what he was thinking, which was: this is nine years of institutional knowledge trapped in a format that will disappear the day you retire.

Instead he spent a weekend configuring a scheduling tool. He imported the department's minimum staffing requirements, certification constraints, leave patterns, and the deputy's own logic — which he extracted, carefully, through three conversations over the following two weeks. He asked the deputy to explain every decision rule embedded in the spreadsheet. He listened. He asked follow-up questions. He wrote everything down.

Then he showed the deputy the tool.

The deputy's first reaction was suspicion. He asked how the system knew about the Tuesday medical appointment. Marcus showed him. He asked what would happen if the system's output conflicted with the deputy's judgment. Marcus said the deputy's judgment would override it, always, and that the system's job was to handle the volume so the deputy's judgment could focus on the parts that actually required it.

The deputy used the tool for the next quarter's schedule. It took him forty minutes instead of three afternoons. He called Marcus into his office afterward.

'You know what I did with the rest of that time?' he said.

Marcus waited.

'I went to Station 4. Just went. No agenda. Stayed two hours.' A pause. 'I haven't done that without a reason in three years.'

Suspicion. Relief. Then the question that opened everything else. Marcus has watched it happen in every department he has worked in since.

The Question in the Apparatus Bay

Dana is not standing in the apparatus bay because ChiefBot surprised her. She is standing there because ChiefBot asked her a question she has been successfully avoiding for eleven years.

If the machine did all of that in four hours — what have I been doing?

It is the right question. It is uncomfortable for a reason.

The administrative burden fire chiefs carry is real. But it is worth asking honestly: how much of it is structurally necessary, how much is inherited habit, and how much exists simply because no one has ever stopped to ask whether it could be done differently? Some administrative work is genuinely inescapable — regulatory requirements, accreditation obligations, legal documentation that must be personally reviewed and approved. That work is not going away.

But some of it persists because it has always been done by the chief, and therefore the assumption is that it must be done by the chief. The policy that could be drafted by a deputy and reviewed in fifteen minutes gets drafted by the chief because that is how it has always worked. The schedule that could be generated by software and approved in five minutes gets built by hand over three afternoons because no one has configured a better system.

And some of it persists because the chief has not had occasion to stop and ask the question. The administrative tide has been high enough, consistently enough, that there has never been a morning like this one — a morning when Dana arrives to find the tide gone out, and stands in the silence of her own apparatus bay, and has to confront what is left on the beach.

She goes back to her desk eventually. Opens the laptop. Begins reviewing the drafted policy updates — the chief still makes the final call, and there are three of them waiting. She reads the first one carefully. It is technically sound. It reflects current NFPA guidance. It is written in language consistent with the department's existing policy format.

She makes two edits. Not because the draft is wrong — because it lacks context that only she has. The new language is correct as a standard; her edit makes it correct for this department, this city, this union contract, this history. The kind of correctness that a system operating without a context document cannot produce and a chief operating without that context cannot protect.

She saves the document. Queues it for review. Moves to the next one.

It takes her thirty-five minutes to review all three. Three weeks of anticipated work, reviewed and finalized in thirty-five minutes.

She looks at her calendar. It is 8:47 a.m. She has the whole day.

She does not know yet what to do with that. But she is beginning to understand that figuring it out is the actual job.

Administration as Identity

There is one more thing worth naming before we move on, because it explains Dana's reaction more fully than simple surprise.

For some chiefs, administrative competence is not just a job function. It is an identity. The chief who can build a flawless ISO package, who knows every policy by heart, who manages the schedule with a precision that borders on art — that chief has built a professional reputation on a particular set of skills, developed over years, refined through iteration, earned through the specific suffering of doing difficult things until they become second nature.

When AI threatens to replicate those skills, what is threatened is not just the work. It is the self-concept.

This is worth taking seriously, not dismissing. Professional identity is real. It is one of the most durable things in a person's interior life, and transitions that touch it require more than a software tutorial. The deputy chief with the color-coded spreadsheet was not being irrational when his first reaction to the scheduling tool was suspicion. He was protecting something that was genuinely his — nine years of accumulated knowledge, encoded in a format that made it visible and legible and proof that he had been paying attention.

Marcus understood this, which is why he spent two weeks extracting the deputy's knowledge before showing him the tool. He was not replacing the deputy's expertise. He was liberating it from the format that had been containing it.

The expertise is still the deputy's. The two hours at Station 4 are still his. The tool handled the volume. The judgment, the institutional knowledge, the relationships, the presence — those are irreducibly human, and they became more available the moment the tool took the volume away.

If you find yourself resistant to AI administrative tools in a way that feels disproportionate — if the resistance is less about the tool and more about what the tool's competence implies about yours — that is information worth sitting with. Not as a criticism. As a diagnosis.

The answer is not to reject the tool. The answer is to build a professional identity that is not contingent on tasks a machine can replicate. Which means understanding, precisely and honestly, what a machine cannot replicate. That is what the rest of this book is about.

Dana is standing in the apparatus bay because, for the first time in eleven years, she does not have an answer to the question of what she is for when the paper tide goes out.

This book is an attempt to give her one. And you one.

FOR THE READER

Three questions. All three deserve an honest answer.

1. If you reclaimed ten hours a week from administrative tasks, what would you do with them? Not what you should do — what you would actually do. If the first honest answer is 'other administrative tasks,' the problem is not the tools. It is the system. And possibly the habit. Dana stood in the apparatus bay for fifteen minutes before she went back to her desk. What she was doing, whether she knew it or not, was trying to locate what she is when the administrative work is gone. That question is worth sitting with before you configure a single tool.

2. Look at your administrative workload honestly: how much of it exists because it must, and how much exists because no one has ever stopped to question it? Inherited administrative burden is common in the fire service. So is the assumption that the chief's job is to absorb it. The deputy's color-coded spreadsheet was not a bad system. It was a good system that had become the wrong vessel for the knowledge it contained. Audit your own spreadsheets — literal and figurative. Ask which ones are protecting something valuable and which ones are just protecting habit.

3. This week — not next month, not after the next budget cycle — pick one recurring administrative task and spend two hours exploring whether an AI tool can handle it. Scheduling, compliance tracking, and policy review cycles are good starting points. The goal is not to replace yourself. The goal is to find out what you are capable of when the paper tide goes out. Dana took thirty-five minutes to review three policy drafts that would have taken her three weeks to produce. What she learned in those thirty-five minutes was not about AI. It was about what her actual job is.

The trucks are still in the bay. The community still needs a chief.

The question is what kind.

Chapter 2

The Data Chief

Scene

The report runs at 2:14 a.m. on a Wednesday.

Nobody asked for it. ChiefBot 9000 runs analyses during low-utilization hours because low-utilization hours are when the processing is fastest. Wednesday at 2:14 a.m. is about as low-utilization as the week gets.

The dataset spans five years. 54,847 incidents. ChiefBot cross-references response times against shift assignments, apparatus positioning, day of week, time of day, weather conditions, call type, and geographic distribution. It flags correlations. It weights them. It builds a model. It does not consider whether the findings will be welcome, or anything else. It processes.

By 2:31 a.m., ChiefBot has identified an anomaly.

Response times to the industrial corridor on the department's eastern boundary run an average of 41 seconds longer on weekday mornings between 6:00 and 8:00 a.m. than at any other time period. The deviation is consistent across all five years in the dataset. It correlates strongly with shift change overlap — a seventeen-minute window during which apparatus positioning is in transition and crew accountability is momentarily distributed.

The corridor contains a chemical storage facility, two manufacturing plants, and a railyard. By call type and occupancy classification, it is the highest-consequence geographic area in the department's response zone.

The deviation has existed for five years.

ChiefBot generates a summary. Three pages. Maps. Charts. A recommended apparatus repositioning protocol for the affected window. Projected response time improvement: 28 seconds average.

It queues the report for Chief Mercer's 0630 review. Then it moves to the next task.

It does not flag this finding as urgent. It does not distinguish between this report and the overtime reconciliation it completes twelve minutes later. Both items are processed. Both are queued. ChiefBot moves on.

Dana reads the report over her first cup of coffee.

She reads the executive summary twice. Then she opens the maps section and sits with it for a long time — the industrial corridor highlighted in amber, the response time deviation color-coded across five years of incidents, the recommended repositioning overlaid on the apparatus coverage grid.

She sets the report down.

The data was always there. All of it. It was always hers, sitting in the records management system, waiting for someone to ask it the right question.

Five years.

She's not sure yet whether she feels relieved that ChiefBot found it or angry that she didn't.

She doesn't immediately decide what to do with it. That's not how she works. She puts the report in the physical folder she keeps for decisions she's not ready to make, the one that sits in the upper right corner of her desk, and she goes about the rest of her morning.

She doesn't go about the rest of her morning. Not easily. She goes through the motions of it. In a meeting with the battalion chiefs at 0900 she catches herself looking at the window facing east — facing the corridor — while someone is talking about training schedules.

After the meeting she drives to the industrial corridor.

She doesn't call it an inspection. She doesn't call it anything. She drives east on Route 9 and takes the industrial access road and parks where she can see the chemical storage facility and the rail yard, and she sits there for twenty minutes watching the morning shift change.

6:47 a.m. Apparatus in transition. A nineteen-minute window during which the nearest unit is seven minutes further from that facility than it would be at any other point in the day.

She's driven this road hundreds of times. She has driven past that facility on the way to incidents and on the way back and on occasions when she had no particular reason to be out here at all except that she likes driving and the eastern end of the response zone has a the unique quality of solitude, industrial solitude, she's always found clarifying.

She never stopped and watched the shift change from this angle.

ChiefBot did not tell her anything she couldn't figured out. It told her something she had not figured out, in five years of driving past it.

She puts the car in drive and heads back to the station.

Story

Every fire department in the country collects data. Most of them are drowning in it.

NERIS reports. CAD records. Apparatus maintenance logs. Training completion databases. Personnel files. Incident after-action reports. ISO documentation. Budget expenditures. Mutual aid requests. Response time logs. Occupancy inspection records. The modern fire department generates more data in a single month than a chief officer could meaningfully review in a year — and yet, in department after department, the dominant mode of decision-making remains experience, instinct, and what worked well last time.

This isn't a criticism, it's the reality of data. Data is only useful when it's organized, analyzed, and interpreted — and until recently, doing that at meaningful scale required either a dedicated analyst, an expensive software platform, or both. Most fire departments have neither. So, the data sits. It gets filed. It satisfies reporting requirements. And then it waits, largely unexamined, for someone to ask the right question.

AI changes that equation. Not by replacing the chief's judgment — but by making the data actually available to it.

What ChiefBot found in the industrial corridor was not magic. It was pattern recognition applied to data that already existed, already belonged to the department, and had been sitting untouched for five years. The insight was always there. What was missing was the capacity to see it.

What AI Actually Does With Fire Service Data

Predictive response analytics — the application at the center of this chapter's scene — identifies patterns in response performance that are invisible to conventional reporting. More importantly, these systems project forward: flagging conditions under which response times are likely to degrade before an incident occurs, rather than after. Departments using them have repositioned apparatus, adjusted staffing windows, and renegotiated mutual aid agreements based on what the data revealed. The decisions were made by humans. The information enabling those decisions came from the machine.

Risk modeling and community vulnerability mapping layers incident history against occupancy data, building age, socioeconomic indicators, and geographic factors to generate community risk profiles that prioritize prevention and inspection resources with a precision that manual assessment cannot approach. A department that knows, specifically and quantifiably, which occupancies in its jurisdiction represent the highest probability of a life-safety incident can deploy its prevention resources accordingly.

Budget forecasting and financial modeling transforms the annual exercise in advocacy-under-uncertainty into a data-backed case that is considerably harder for municipal financial officers to dismiss. AI tools that analyze multi-year expense patterns, apparatus age and use trajectories, staffing trends, and call volume growth produce projections with a level of specificity that changes the conversation — not just the numbers.

After-action analysis may be the most underutilized application. After-action reviews are standard practice. The combination and assessment of patterns across dozens or hundreds of them over time is not. AI tools can process large volumes of unstructured after-action data, identify recurring themes, flag systemic issues, and surface training needs that individual reviews miss. What looks like three separate near-misses over eighteen months may, in aggregate, reveal a single procedural gap that no individual review identified.

The Decision You Made Last Week

Think about a significant operational or administrative decision you made in the last thirty days. A staffing call. A resource deployment. A budget prioritization. A policy change. A training focus area.

Now ask: what data informed that decision?

For most chiefs, the honest answer involves some combination of experience, professional judgment, input from trusted subordinates, and whatever information happened to be readily available. That is not negligence. That is how fire service leadership has functioned for generations, and it has produced many good decisions by skilled people.

But it's also produced decisions that better data would have changed. Apparatus positioned based on historical patterns that have shifted. Training prioritized based on what the chief remembers needing, not what current incident data suggests is lacking. Inspection resources deployed based on familiarity rather than risk modeling.

Marcus Webb directly encountered this at his second department — a mid-sized suburban department with solid operational performance and a prevention program that had not materially changed in eight years. When Marcus ran a basic risk analysis against five years of incident and occupancy data, the results were illuminating and uncomfortable: the department's prevention resources were heavily concentrated in commercial occupancies, while the data showed that residential fires — particularly in a cluster of aging housing stock in the department's northwest quadrant — represented the highest frequency and severity of life-safety incidents.

The prevention program was well-run. It was aimed at the wrong target.

Marcus remembers sitting with the battalion chief who ran the prevention division — a twenty-year veteran who had built the program from a one-person operation into something the department was genuinely proud of — and showing him the data. He waited for the defensiveness. It didn't come. The battalion chief looked at the maps for a long time. Then he said: 'We've been proud of the wrong thing.'

The program shifted. Prevention resources reoriented toward residential occupancies in the northwest quadrant. Inspection frequency adjusted based on risk modeling rather than historical rotation. Community outreach prioritized by vulnerability profile rather than by which neighborhoods the department already had relationships with.

Residential fire incidents in the northwest quadrant declined measurably over the following two years.

Nobody made a bad decision. Nobody had been negligent. The data just hadn't been looked at that way. The battalion chief hadn't been wrong to build what he built. He had been working with the information he had. Better information changed what he built.

What the Data Cannot Tell You

Data is not judgment. It's input to judgment.

The pattern ChiefBot identified — the 41-second response time gap in the industrial corridor — is actionable information. But acting on it requires considerably more than the data alone provides.

It requires a chief who understands the implications of labor relations in changing shifts to adjust for overlap to reduce response lag. The shift change protocol is not arbitrary; it exists because it was negotiated, or because it reflects a longstanding operational practice that crews have built their routines around, or both. Changing it means understanding the contract, understanding the informal norms that have grown up around the formal policy, and anticipating the resistance that will come from the people most directly affected.

It requires knowing which captains will resist the apparatus repositioning and why. Not because resistance is always wrong — sometimes the people closest to the operational reality have information the data cannot capture — but because understanding the nature of the resistance is how you distinguish legitimate operational concern from institutional inertia. You cannot make that distinction from a spreadsheet.

It requires the political standing with the union to propose the change without triggering a grievance. A repositioning protocol that's operationally correct but procedurally handled badly can cost more in labor relations than it recovers in response time. The chief who has spent years building credibility with union leadership can propose this conversation. The chief who has not cannot, regardless of what the data says.

It requires the ability to make the case to city council that the chemical storage facility in that corridor justifies the operational adjustment — in language that a council member without fire service experience can understand, using risk framing that resonates with elected officials whose primary concern is liability, budget, and relations with voters.

Dana drives back to the station understanding all of this. She has the data. She does not yet have a plan. Between the data and the plan there's a map of her organization — its relationships, its history, its informal structures, its political landscape — that ChiefBot doesn't have and cannot have. That map lives inside her. She built it over eleven years.

ChiefBot found the pattern. It takes a chief to do anything about it.

The chief who ignores the data is flying blind. The chief who trusts only the data is delegating judgment to a system that has none. The chief who uses the data to sharpen their own thinking — and then applies the full weight of their experience, relationships, and institutional knowledge to act on it — that chief is doing something no algorithm can replicate.

Data tells you what is happening. It rarely tells you what to do about it. That is still your job.

Dana doesn't call a meeting about the industrial corridor that week. She has two conversations — informal, no agenda, the kind she's learned to have before the kind with agendas. One with the battalion chief whose district includes the corridor. One with the union president, over coffee, about nothing in particular.

She's building the map. She is figuring out where the resistance will come from and what it will be made of. She's doing the work that comes before the work, the work that doesn't show up in any data system and never will.

Three weeks later she will call the meeting.

ChiefBot found the pattern in seventeen minutes. Dana's preparation took three weeks. Both were necessary. Only one of them could have been done by a machine.

FOR THE READER

Three questions. The same discipline applies.

1. What data does your department collect that you've never seriously analyzed? Pull up your CAD data export, your records management system, or your training records database and look at it — not for a specific purpose, but with genuine curiosity. Ask: what patterns might be here that I've never thought to look for? The industrial corridor pattern wasn't hidden. It was unasked. If you don't know where to start, start with response times by geographic zone and time of day. The pattern may not be dramatic. It might also be exactly what ChiefBot found for Dana.

2. Identify one decision you made in the last ninety days that was based primarily on experience or instinct. Not a bad decision — just one where data was not a significant input. What data exists in your department that could have informed that decision? Was it available? Was it accessible? Was it organized in a way that would have been useful? The gap between the data you have and the data you use is where AI works. The point is not that instinct is wrong. The point is that instinct informed by data is better than instinct alone.

3. When you find a pattern — and you will, if you look — resist the instinct to call a meeting immediately. Do what Dana did: have two conversations first. One with the person closest to the operational reality the data has identified. One with the person whose buy-in will determine whether you can act on what you found. The data is the beginning of the work, not the end of it. The work that follows is irreducibly human, and it's where your value lives as a chief.

The pattern was always in the data.

The question is whether you are looking.

Chapter 3

The Training Chief

Scene

The request comes from the deputy chief on a Tuesday morning: generate a training needs assessment for all sworn personnel, prioritized by individual performance gaps, produce recommended curricula for the next twelve months.

ChiefBot 9000 acknowledges the request in 0.4 seconds.

By Thursday afternoon, the assessment is complete.

Ninety-four individual training profiles. One per sworn member. ChiefBot has pulled five years of incident response data and cross-referenced it against individual personnel records — who was on which apparatus, in which role, during which call types. It has analyzed after-action reports for recurring deficiencies, weighted by incident severity and frequency. It has reviewed every training completion record, mapped certifications against current NFPA standards, and flagged seventeen members whose credentials will lapse within the next six months. It has cross-referenced injury reports against training histories and identified three skill domains where members with fewer than five years of service show statistically elevated injury rates during specific incident types.

Each profile identifies the member's demonstrated strengths, their documented gaps, the incident types where those gaps carry the highest operational risk, and a recommended sequence of interventions with timelines.

Engine Captain Delgado: twelve years of service, consistently strong on structural fire operations, recurring deficiencies in hazmat first-responder operations — specifically initial identification and assessment. Recommended: hazmat awareness refresher, scenario-based tabletop, live evolution with evaluation. Timeline: ninety days.

Firefighter Okonkwo: three years of service, above-average performance on EMS assist calls, below-average on high-rise operations — a gap consistent with the department's limited high-rise incident history and the member's training cohort. Recommended: high-rise familiarization, stairwell operations drill, emergency elevator operations certification. Timeline: sixty days.

ChiefBot queues the report and moves on to the next task.

It doesn't register that the ninety-fourth profile — the last one in the document — belongs to the chief.

Dana opens the file on a Friday morning. She scrolls through the first several profiles with something between relief and unease. Immediately she can see the assessments are accurate. They're organized the way she would've organized them if she'd been given the time to individually analyze each firefighter.

She reaches her own profile and stops.

ChiefBot has assessed the chief. Documented strengths. Documented gaps. A recommended sequence of development interventions with a timeline.

She reads it once. Then again.

It wasn't wrong.

She closes the laptop. Not because she's done — because she needs a minute that isn't in front of a screen.

Eleven years as a fire chief. She has reviewed hundreds of performance assessments in that time. She's had the developmental conversations, delivered the honest feedback, sat with members who were struggling and members who did not know they were struggling and members who knew and didn't want to talk about it. She has been the person on the other side of the table her whole leadership career.

She has never been on this side of it.

The profile identifies three areas. The first two she already knew, more or less — gaps she has been aware of and managing around, the way a person manages around a familiar limitation rather than confronting it directly. The third one she didn't know. Five years of incident data, cross-referenced against the decisions she made in her role as IC, surfacing a pattern she had not seen in herself.

She sits with that for a while. The specific discomfort of accurate information about yourself that you didn't ask for and cannot dispute.

Then she does something she will think about later as the right call but could easily not have done: she picks up the phone and calls the division chief of training.

'I need you to look at something,' she says. 'Come to my office when you get in.'

She doesn't tell him what it is. She's not ready to say it out loud yet.

Story

Training in the fire service is personal. That is both its greatest strength and its most persistent limitation.

The best training officers most of us have encountered shared certain qualities — they knew their people, they understood the gap between what the SOG said and what actually happened on the fireground, and they designed evolutions that challenged without embarrassing, stretched without breaking, and built competence through repetition and honest feedback. They were students of their craft and students of their people simultaneously. That kind of training is irreplaceable. It is also rare, inconsistent, and extraordinarily difficult to scale.

The structural reality of fire service training is that most departments design their programs based on some combination of regulatory requirements, available budget, instructor availability, and the training officer's professional instincts. All of those inputs are legitimate. None of them is the same as a systematic, data-driven analysis of where individual members are actually falling short and why.

The result are a training programs that are often well-designed in the aggregate and poorly targeted at the individual. Everyone runs the same evolutions. The twelve-year veteran with a specific procedural gap in one narrow skill domain sits through the same awareness-level refresher as the probationary firefighter encountering the material for the first time. The system treats everyone the same because it's never had an efficient mechanism to do otherwise.

This is where AI changes the equation — not by replacing the training officer's craft, but by giving that craft a precision it has never had before.

What a Real AI-Assisted Training Needs Assessment Looks Like

Step one: gather your data sources.
A meaningful training needs assessment draws on at least four data streams that most departments already have but rarely connect: incident response records from your CAD or RMS system, individual training completion records, certification status by member, and after-action reports. You don't need all four to start — two will tell you more than none. These data streams exist in isolation in most departments, filed in separate systems for separate purposes, and the usefulness of their synthesized data never happens because no one has the time or tools to do it manually at scale.

Step two: define what you are looking for.
A training needs assessment is only as useful as the questions it's designed to answer. Are you looking for individual skill gaps, program gaps, or both? Are you focused on a specific incident type, how a hiring class has evolved since being hired 10 years ago, or the department as a whole? This step cannot be automated. It requires human judgment about what the department's current risk profile demands. The AI do'sn't decide what questions to ask. You do. The AI helps you answer them.

Step three: run the analysis.
Once the questions are defined and the data organized, the analytical work begins. You're looking for correlations — between incident types and individual performance, between training completion and operational outcomes, between certification status and deployment patterns. You're looking for frequency — which skill domains appear repeatedly in after-action deficiency notes. You're looking for gaps — between what your training calendar emphasizes and what your incident data says you need.

Step four: translate findings into action.
The output of a training needs assessment is not a report. It's a set of decisions waiting to be made. Which gaps are most urgent? Which members

need individual attention, and which gaps indicate a program-level response? These decisions require a human with operational knowledge, relationships with the personnel involved, and judgment about what the department can realistically accomplish. The assessment tells you the situation. You decide what to do about it.

Step five: evaluate and iterate.
After interventions are delivered, the same data streams that identified the gaps will tell you whether they're closed. This feedback loop is what transforms a training program from a series of events into a learning system. AI makes it feasible. The system that identified the gap can monitor for the improvement continuously and without additional effort.

The Problem With Training Everyone the Same

The assumption that everyone needs the same training at the same time is a legacy of resource scarcity and administrative convenience, not a pedagogical principle. It persists because individualized training has historically been prohibitively expensive to design and nearly impossible to schedule at scale. AI reduces those constraints significantly — and in doing so, makes a question unavoidable: if we can now design training that meets people where they actually are, why are we still designing training that meets everyone in the same imaginary middle?

The Training Officer who asks that question — and uses available data and tools to answer it rigorously — is doing something fundamentally different from the one who schedules the same calendar as last year with minor modifications. The former is building a learning organization. The latter is maintaining a compliance record. Both keep the boxes checked. Only one develops the people.

What AI Cannot Design

The division chief of training arrives at Dana's office at 0830. She has the laptop open. She turns it toward him without saying anything.

He reads the chief's profile. She watches him read it. He's careful not to react, which is its own kind of reaction.

'Where did this come from?' he asks.

'Same place the other ninety-three came from,' she says. 'ChiefBot.'

He reads it again. 'The third one,' he says. Not a question.

'Yeah.'

He sits back. There's a long pause, the kind that exists only between people who have worked together long enough to not need to fill every silence.

'Do you want me to tell you it's wrong?' he asks.

'Is it?'

He doesn't answer immediately. Which is answer enough.

'What do you need from me?' he says.

'I need you to treat my profile the way you'd treat any other profile. Design something. We'll figure out how to run it without making it a department announcement.'

He nods. He closes the laptop. He gets up to leave.

'Dana.'

She looks up.

'The fact that you called me in here is the thing that matters. Most chiefs would have filed it.'

She thinks about that for a while after he leaves.

Training is a relationship. The best training officers are skilled readers of people, not just designers of curriculum. They understand that the same feedback lands differently depending on who delivers it, under what circumstances, in front of whom, and with what relationship behind it. They know which members need to be challenged publicly to rise to it, and which need a private conversation to hear something difficult without becoming defensive. That knowledge comes from relationship, not data.

Psychological safety is not a soft concept. It's a measurable organizational condition with direct operational consequences. Research on high-reliability organizations — military aviation, nuclear power, surgical teams — consistently identifies psychological safety as the variable that separates organizations that learn from near-misses from those that repeat them. In its presence, people surface problems early, engage honestly with feedback, and use training as a genuine opportunity to improve. In its absence, people hide gaps, half-mimic competence they don't have, and treat training as a compliance exercise to complete rather than an opportunity to use.

The fire service has a complicated relationship with psychological safety. The culture of operational confidence — the expectation that you know your job and minimize the amount of uncertainty you advertise — serves real purposes on the fireground. Applied to the training environment without modification, it produces members who will not admit what they don't know, will not ask questions that reveal and address knowledge or skill gaps, and will not engage honestly with feedback that challenges their self-assessment.

Consider what that looks like in practice: a training evolution where an experienced firefighter makes a procedural error in front of peers. In a high-psychological-safety environment, that firefighter acknowledges the error, the training officer uses it as a teaching moment, and the crew learns from what happened. In a low-psychological-safety environment, the firefighter minimizes the error, the training officer moves on rather than risk a confrontation, and the crew learns that errors are not addressed,and deviations from the standard are normalized here. That training generates a completion record. It does not generate learning.

The training program generates completion records without generating competence. And the next time that firefighter encounters the same situation on an actual incident, the gap is still there — unaddressed, unacknowledged, and now consequential.

ChiefBot identified the gap. It can't build the culture in which closing the gap is safe. That's the training officer's job, and it is unequivocally human.

The most effective use of AI-generated training profiles is a preparation for human conversations, not a substitute for them. The training officer who reviews ChiefBot's assessment of a member before sitting down with that person is better prepared for that conversation than one who goes in without it. They know what the data shows. They can ask better questions. They can listen for the context the data cannot capture — the reason behind the gap, the experience that shaped the pattern, the constraint that doesn't surface in the training and incident records.

The assessment tells you where the gap is. The relationship is how you close it.

Dana distributes the training profiles to her battalion chiefs and division chiefs the following Monday. She doesn't email them. She meets with each one individually — four conversations over two days. She explains what ChiefBot generated and how. She tells them the profiles are starting points, not verdicts. She tells them to use the profiles to prepare for conversations with their people, not to replace those conversations.

She doesn't tell them about her own profile. Not yet. She's still working out how to handle that part in a way that models what she's asking of them without making it about her.

That's the question ChiefBot cannot answer. How do you use accurate information about yourself as a leader, in a way that builds rather than undermines the culture you are trying to create?

There's no algorithm for that.

FOR THE READER

Three questions. The same rules apply.

1. Pull your last twelve months of incident data and your last twelve months of training records side by side. Ask: is the training we delivered connected to what our incidents are telling us we need? If the answer requires more than thirty seconds of thought, the connection is probably weaker than it should be. Audit the gap. The point is not to find failure — it's to find the delta between what you are teaching and what the work is actually demanding. That delta is where the next twelve months of training design should start.

2. Identify one training gap in your department that you suspect exists but cannot currently prove — a skill domain or an incident type where performance feels inconsistent, but the data has never been organized well enough to confirm. What data do you already have that could tell you whether the gap is real or simply a subjective viewpoint? The goal is to move from suspicion to evidence, and

from evidence to targeted intervention. ChiefBot found Dana's gap in the industrial corridor because someone asked the right question. Ask the question.

3. Read your own profile. Not a ChiefBot profile — you may not have one yet. But you have a sense of where your developmental edges are. Most leaders do if they're honest with themselves in a quiet moment. Name three. Write them down. Then ask: what would I need to do to close one of them this year? Not whether it will be comfortable. Whether it will be worth it. Dana called the division chief. The fact that she called him is the thing that matters. Most chiefs would have filed it away.

The data knows what your people need.

The relationship is how you do something about it.

Chapter 4

The Strategic Chief

Scene

The city manager's request arrives on a Friday afternoon: a five-year strategic plan for the fire department, due in sixty days, suitable for presentation to the city council and the municipal budget committee.

ChiefBot 9000 logs the request at 4:47 p.m.

By the following Wednesday, the plan is complete.

Ninety-four pages. A community risk assessment built from five years of incident data, census demographics, building permit activity, and infrastructure age maps pulled from the city's GIS system. A staffing analysis projecting call volume growth against current personnel levels, with a recommended hiring timeline and cost projections broken into fiscal years. An apparatus replacement schedule derived from manufacturer specifications, maintenance cost trajectories, and industry benchmarks. A facilities assessment cross-referencing station locations against current and projected population density, with three proposed coverage optimization scenarios. A grant opportunity inventory — seventeen federal and state programs for which the department currently qualifies — with eligibility windows, application requirements, and projected award amounts. A community risk reduction strategy built around the department's top five incident categories by frequency and severity. An executive summary. A budget narrative. A slide deck suitable for council presentation.

ChiefBot flags two items requiring human input: the department's stated values and strategic priorities, which ChiefBot has drafted based on IAFC model language but flagged as requiring chief officer review and personalization; and the community engagement strategy, which ChiefBot has outlined but noted cannot be finalized without the chief's assessment of existing relationships and political context.

It queues the plan and moves to the next task.

Dana opens the document on Thursday morning. She reads the executive summary. Then the staffing analysis. Then the apparatus schedule.

She pulls up the grant inventory and scrolls through it slowly.

There are seventeen programs listed. She's aware of four of them. She has applied to two. She's been meaning to look into the others for three years. There's never enough time.

She does not feel gratitude, exactly. She feels the specific discomfort of someone who's just discovered that the thing they told themselves they would get to when they had more time was actually achievable the whole time — it just required a different tool.

Three years. Thirteen programs she qualified for and never had time to investigate. She scrolls back to the top and reads the eligibility window for the first unfamiliar one.

The window closed eight months ago.

She closes the laptop. Sits for a moment. Then opens it again and moves to the next one.

Story

Strategic planning occupies a complicated position in fire service leadership. It's universally acknowledged as essential — every accreditation program, every municipal governance structure expects the fire chief to articulate a vision, set priorities, and chart a multi-year course. And it's the function most reliably crowded out by everything else. Operational demands. Administrative burden. The relentless pressure of managing a twenty-four-hour emergency response organization.

The result, in department after department, is strategic plans drafted under deadline pressure, built on incomplete data, and then filed — genuinely filed, in a drawer or a shared drive — until the next accreditation cycle or the next city manager asks for it. The plan exists. It's rarely alive.

AI removes the most significant structural barrier to genuine strategic planning: the volume of data gathering, synthesis, and document generation that consumes the available time before the thinking can even begin. When ChiefBot produces a 94-page plan in five days, it's not doing the strategic thinking. It's clearing the desk so the strategic thinking can happen.

What ChiefBot Actually Built

Community risk assessment.

A thorough risk assessment draws from incident history, demographic data, occupancy classifications, the condition of infrastructure, geographic and environmental factors, and trend analysis over time. AI tools can ingest these data streams — many publicly available through municipal GIS systems, census databases, and state fire reporting platforms — and produce a structured risk profile that would require weeks of manual analysis. The resulting document is not just a planning input. It's a data-backed foundation for every resource and priority decision within the strategic plan.

Staffing analysis and workforce planning.
Most experienced chiefs have a strong instinct about whether their department is adequately staffed. Translating that instinct into a projection that accounts for call volume growth, retirement timelines, minimum staffing requirements, and budget constraints — presented in a format that a municipal finance department finds credible — is a different task. AI tools model staffing scenarios across multiple variables simultaneously, producing projections that are both analytically rigorous and narratively coherent for non-fire-service audiences.

Apparatus lifecycle management.
Apparatus replacement decisions are often made reactively rather than strategically. AI-assisted lifecycle modeling looks at the full cost trajectory of an apparatus — maintenance history, manufacturer specifications, peer and neighboring department replacement benchmarks, parts availability trends — and produces replacement schedules that optimize both operational readiness and budget predictability. The chief who walks into a capital budget conversation with a ten-year apparatus replacement schedule supported by cost modeling is in a fundamentally different position than the one presenting a list of trucks that are getting old.

Grant research, identification, and application development.
The federal and state grant landscape available to fire departments is extensive, frequently updated, and requires sustained attention to navigate effectively. FEMA's Assistance to Firefighters Grant program alone distributes hundreds of millions of dollars annually. The Staffing for Adequate Fire and Emergency Response program funds hiring. The Fire Prevention and Safety grant funds community risk reduction and research. Most departments pursue the grants they already know about and miss the ones they have never heard of.

AI tools can continuously monitor grant databases, match department characteristics against eligibility criteria, and surface opportunities with sufficient lead time to prepare competitive applications. More directly, AI can assist with the application itself. Each scored section of an AFG application — the project description, the financial need narrative, the effect on daily operations, the cost-effectiveness justification — requires clear, specific, well-organized prose that speaks to FEMA's stated priorities for that grant cycle.

The chief who provides an AI tool with accurate information about their department and asks it to draft the narrative sections of an AFG application will receive a starting document that's more clearly organized and more compellingly written than most first drafts produced without it. The chief then reviews it for accuracy, personalizes it with department-specific detail, and verifies it against the current year's Notice of Funding Opportunity before submission. That last step is not optional. FEMA's priorities shift annually, and an AI tool's knowledge has a cutoff date. The chief who submits an AI-assisted application without verifying it against the current NOFO is not using AI well. The chief who uses AI to produce a strong draft and then applies their own judgment to refine and verify it *is* doing exactly what this book advocates.

Is using AI for grant writing ethical? Yes — with the same clarity that makes hiring a grant writer ethical, using a template ethical, or having a deputy chief edit your narrative ethical. What matters is that the content is accurate, the need is genuine, and the chief stands behind every word of the submission.

Dana spends Friday afternoon with the grant inventory. She opens a new document alongside it and starts a list: which programs have upcoming windows, which require coordination with other city departments, which she can pursue independently. It's not a plan yet. It's triage.

She finds two programs with application windows in the next four months that the department clearly qualifies for and has never applied to. One is for fire prevention and community education — a program she has wanted to expand for two years without the budget to do it. The other is a SAFER grant that would fund two additional firefighter positions, which she has needed but couldn't figure out how to ask for without the data to back it up.

She has the data now. ChiefBot pulled it, formatted it, and embedded it in a draft narrative that's seventy percent of the way to submittable.

She prints the grant inventory. She carries it home.

Three years of meaning to get to this. Five days for the machine to clear the deck.

The Strategy the Algorithm Cannot Write

ChiefBot flagged two items requiring human input, and the specificity of these flags is important.

The first was the department's stated values and strategic priorities. Not because ChiefBot couldn't draft them — it did, using IAFC model language — but because it recognized, correctly, that values and priorities are not data outputs. They're leadership choices. They reflect what the chief believes the department is for, what kind of organization it aspires to be, and what it's willing to trade off to get there. A strategic plan whose values were written by an algorithm and never meaningfully reviewed by the chief is not a strategic plan. It's a formatted document.

The second flag was the community engagement strategy. ChiefBot could outline one. It could not write one, because community engagement is a relationship function — it depends on knowing which neighborhood associations have a history with the department, which council members have been allies and which skeptics, which community organizations have credibility with the populations the department most needs to reach.

Marcus Webb learned this at his third department. He inherited a plan drafted by a consultant, adopted by the city council, and promptly forgotten by everyone in the building. It described a vision for the organization that no one recognized as their own — because no one had been asked.

Marcus spent his first ninety days listening. He met with every company officer individually — not to discuss the plan, but to understand what each of them thought the department's most significant challenges were. Some of those conversations were fifteen minutes. Some ran an hour. One captain — a twenty-six-year veteran three years from retirement, the kind of officer who has stopped saying what he thinks because it has never mattered — talked for ninety minutes. Marcus didn't check his phone. He took notes.

What he heard, across forty-one conversations, was a department that knew exactly what it needed and had completely stopped expecting to be asked.

Marcus held a series of structured conversations with his senior leadership team about values — not the laminated-card version, but the actual operating values that governed how they made decisions when no policy covered the situation. He asked them to describe a moment when someone in the department had done something that made them proud — not a heroic fireground moment, but a leadership moment. The responses told him more about the department's actual values than any document he had inherited.

When Marcus then used AI tools to rebuild the data foundation of the plan, he brought the company officers back into the room before the priorities were set. He presented the data. He explained what it showed. And then he asked the question most strategic planning processes skip: given what the data tells us about where we are, what do we believe we are obligated to do about it?

The conversation that followed was the most productive planning session Marcus had experienced in twenty-two years. Not because the officers agreed on everything — they didn't. But because they were reasoning from the same data, toward the same values, with a shared understanding of what the plan was supposed to accomplish. The disagreements were productive. The decisions were owned.

The plan that resulted was sixty-seven pages. It was adopted by the city council six weeks later. Company officers cited it in budget conversations. Training priorities were explicitly linked to it. When a staffing shortage emerged in year two, the plan provided the framework for how the department responded.

AI cleared the desk. Marcus did the work that mattered.

Strategy Requires Adaptability

A strategic plan is not a prediction. It is a commitment to a direction, made with the best available information, that must be revisited and revised as circumstances change. The chief who treats a five-year plan as a fixed document — who follows it regardless of what the environment is doing — is not being strategic. They're being rigid.

The most common failure mode is not a bad plan. It's a plan that's never revisited. It's filed with appropriate ceremony, distributed to all the right people, referenced in council presentations, and then left untouched until the next accreditation cycle. Meanwhile the community changes. Call volumes shift. The budget environment evolves. The plan — accurate when written — drifts further from the reality it was meant to guide.

AI tools that continuously update risk assessments, monitor grant landscapes, and track performance metrics make adaptive strategy more achievable than at any previous point in the profession's history. The question is not whether the tools exist. It is whether the chief builds the habit of looking at them.

Dana schedules the plan's first quarterly review before she finishes reading it. Not because she has figured out the adaptability question. Because she's learned, in the past several weeks, that the tools are only as useful as the habits she builds around them. ChiefBot will update the risk assessment. The grant inventory will surface new opportunities. The staffing projections will shift as the budget environment shifts.

Someone has to decide what to do about all of it.

It's the starting line. She's been looking for the starting line for three years.

FOR THE READER

Three questions. The same discipline applies.

1. This month, use an AI tool to draft one element of your department's strategic plan — not the whole document, just one component. The grant opportunity inventory is the most immediately valuable starting point: describe your department's size, staffing model, apparatus inventory, and current service challenges to a well-prompted AI tool and ask it to identify federal and state grant programs for which you currently qualify, along with application windows and projected award amounts. Then check how many of those programs you were already tracking. The gap is the number that matters. Dana found thirteen programs she qualified for and had never pursued. One had a window closing in thirty days.

2. Identify one strategic decision your department made in the last two years that was based primarily on instinct, precedent, or political pressure rather than data analysis. What data existed, or could have been gathered, that would have informed that decision differently? This is not an exercise in second-guessing. It's an exercise in understanding the gap between the strategic planning you've been doing and the strategic planning you *could* be doing. The point is not that instinct is wrong. The point is that instinct supported by a 94-page data analysis is a different kind of instinct.

3. Pull your department's current strategic plan and map it against your current community risk profile. Are the priorities still the right priorities? Are the assumptions still accurate? If the plan is more than eighteen months old and has not been formally reviewed in that time, schedule the review before you close this book. The plan that doesn't get revisited is not a plan. It's a document. Dana scheduled her first quarterly review before she finished reading the plan. That's the habit. Build it before you need it.

A machine can build the foundation, outline, or architecture of a strategy.

Only a leader can make it live.

PART TWO

What AI Cannot Do

Chapter 5

The 2 AM Call

This chapter addresses some of the hardest moments in fire service leadership — including a line-of-duty death, mass casualty incidents, and firefighter suicide. It approaches them directly, because the fire service deserves no less.

Scene

The alert comes in at 2:11 a.m.

Engine 7 is on scene at a residential structure fire, fully involved. The mayday is transmitted at 2:11:43. Firefighter down. Interior crew. The rapid intervention team deploys. The incident commander requests mutual aid. The operations section expands.

ChiefBot 9000 receives the alert simultaneously with the duty officer and begins processing.

By 2:19 a.m. — eight minutes after the mayday — ChiefBot has cross-referenced the crew manifest against personnel records, identified the member, located next-of-kin contact information, and generated a notification protocol document consistent with department policy. It's drafted a next-of-kin notification script. It's identified the chaplain on call and the peer support coordinator and generated contact information for both. It has flagged the department's critical incident stress management protocol and queued a summary for the incident commander. It already began drafting a preliminary media statement — factual, appropriately sparse, pending confirmation — and identified the public information officer's after-hours contact. It opened a critical incident documentation file and began logging the timeline.

At 2:24 a.m., the incident commander confirms: Firefighter James Okafor, eleven years of service, is deceased.

ChiefBot updates the documentation file. It flags the next-of-kin notification as the immediate priority action item. It generates a checklist of required notifications — family, union, department leadership, state fire marshal — with applicable timeframes and responsible parties assigned.

The notification script is ready. It's professional, accurate. It contains the right words in the right order.

It's waiting for someone to deliver it.

Dana's phone rings at 2:26 a.m. She is awake before the second ring. She has been a fire chief long enough that a call at this hour requires no explanation.

She listens. She says very little. She gets dressed in the dark, moving by memory through a room she has navigated for eleven years.

The notification script is in her inbox. Her phone shows the alert from ChiefBot. She does not open it.

She picks up her keys.

She drives across the city at 2:30 in the morning with the radio off.

She's made this drive before — not this exact drive, not to this family, but this drive. Every fire chief of any tenure has made some version of it. The middle-of-the-night crossing of a city that does not know yet what has just happened in it. The houses dark, the streets empty, the ordinary silence of a Tuesday morning that's about to become something else for one family, irrevocably.

She thinks about James Okafor. Eleven years. She knows his face and his voice and the specific way he moved through a building on a search. She knows he has two daughters. She knows his wife's name is Amara. She learned these things not from a personnel file but from being present — from the graduations and the promotions and the department cookouts and the times she stood next to him in apparatus bays and said nothing in particular. From being chief of this department, which means being present in the life of this department, which sometimes means knowing the things about your people that do not appear in any system.

ChiefBot knows his personnel record. It knows his next-of-kin contact. It does not know that his older daughter just started the second grade. Dana knows that because she asked about her at the last department picnic, and James told her, and she remembered.

She parks in front of a house that doesn't know yet.

She sits in the car for thirty seconds. Not to compose herself — she's beyond that kind of composure. To understand, one more time, what she is about to do and why she is the one who has to do it.

Then she gets out of the car and walks to the door.

Story

There is no gentle way into this chapter.

When a firefighter dies in the line of duty, the administrative response — the notifications, the documentation, the media statement, the union coordination, the NIOSH investigation and reporting, the benefit processing — is real and necessary and time-sensitive. ChiefBot can handle most of it competently. In fact, ChiefBot can handle most of it better than a chief officer who is simultaneously managing an active incident scene, notifying department leadership, and carrying the weight of what has just happened to one of their people.

This is not a small thing. Administrative failure in the aftermath of a line-of-duty death compounds the trauma of it. Families who receive inconsistent information, or who are not notified before the news breaks publicly, or who navigate a benefit process that is disorganized and slow — those families carry an additional wound that has nothing to do with the fire and everything to do with how the institution responded. AI that manages the administrative aftermath of an LODD with precision and consistency is genuinely valuable. It should be used.

But the administrative response is not the leadership response. And confusing the two is a failure of understanding about what crisis leadership actually is.

Dana drives to the Okafor family's home at 2:30 in the morning because she is the chief of this department, James Okafor was her firefighter, and the family deserves to hear what happened from a human being who knew him — not a phone call, not a script, not a process. She goes because presence is not a policy. It's a choice. And in the fire service, it's one of the most consequential choices a chief makes.

That choice cannot be automated. It cannot be optimized. It cannot be delegated to a system that processes information without feeling the weight of it.

The Spectrum of Crisis

A line-of-duty death is the most acute form of organizational crisis a fire chief will face. But the principles that govern leadership through an LODD apply across a broader spectrum of critical events.

Serious line-of-duty injury carries many of the same leadership demands as a fatality, compressed into a more ambiguous and often more prolonged situation. The firefighter who survives a critical injury but faces months of recovery, uncertain return to duty, and the psychological aftermath of trauma needs a chief who shows up — at the hospital, at the home, at the rehabilitation facility — not a chief who ensures the worker's compensation paperwork is filed on time. Both matter. Only one requires a human being.

Mass casualty incidents place a different kind of demand on the chief: the management of organizational trauma at scale, across an entire shift or department, while simultaneously managing the operational and administrative aftermath of a major event. The firefighters who worked that scene will carry it. Some will need professional support. Some will never ask for it. The chief who creates conditions in which asking is possible — who names what happened, makes space for the response, doesn't rush the organization back to normal before it is ready — is performing a leadership function that has no algorithmic equivalent.

Firefighter suicide is the crisis the fire service has been least prepared to address — and the one it most urgently needs to confront. The Ruderman Family Foundation's white paper on mental health and first responders, along with subsequent research by the Firefighter Behavioral Health Alliance (FBHA), has established that firefighters are more likely to die by suicide than in the line of duty. In 2024, the FBHA documented 112 confirmed firefighter suicides. By the organization's own estimate, confirmed cases represent only about 40 percent of actual suicides, meaning the true number is significantly higher. Each year, more firefighters die by suicide than from all other line-of-duty causes combined. The leadership response to a firefighter suicide — how the chief communicates about it, whether they name it directly or obscure it, how they support the surviving family and the surviving crew, what organizational signals they send about mental health in the days and weeks that follow — has measurable consequences for the department's culture and for the safety of the people in it. This is neither administrative nor policy work. It is the most personal and most consequential leadership work a chief can do, and it requires a human being with courage, presence, and genuine understanding of the people they lead.

In each of these scenarios, AI can support the administrative and logistical response with competence that frees the chief to focus on what only the chief can do. The documentation, the notifications, the scheduling of critical incident debriefs, the coordination of peer support resources — ChiefBot handles these reliably. What ChiefBot cannot do is sit with a family at 3 a.m. It cannot stand in front of a crew that worked a mass casualty and find the words that honor what they did and acknowledge what it cost them. It cannot call a firefighter's spouse the day after a colleague's suicide to ask how they are doing — and mean it.

Presence as a Leadership Function

The fire service has a functional definition of leadership that centers on decision-making, resource management, and operational command. These are real and important competencies. They are also insufficient.

The chief who is exceptional at incident command and invisible during organizational crisis is not a complete leader. The chief who manages the administrative aftermath of a LODD flawlessly but sends a deputy to represent the department at the funeral is not a complete leader. The chief who processes a firefighter's suicide through the department's EAP referral protocol without ever personally reaching out to the affected crew is not a complete leader.

Presence — physical, emotional, sustained — is a leadership function. Not a soft one. A function. It produces measurable outcomes: trust, cohesion, psychological safety, organizational resilience. Departments whose chiefs show up when it is hard to retain people longer, recover from trauma faster, and build the kind of culture where members feel safe enough to ask for help before they reach crisis. The return on the investment of presence is not always visible in the short term. It's always real.

AI cannot be present. It can simulate availability — generate check-in messages, flag members whose performance metrics suggest elevated stress, schedule peer support contacts. These are useful. They are not the same as a chief who walks into a station three days after a difficult call and asks the crew, without an agenda, how they're doing.

Dana leaves the Okafor home at 4:47 a.m. She doesn't go back to the station. She drives to Station 7 — the house Engine 7 belongs to, where the crew that was on scene tonight will come back to when the incident is cleared.

She sits in the watch office until they return. She doesn't have anything prepared to say. She's learned, over eleven years, that prepared things are usually the wrong things in moments like this one. She's there because they need to see that she came. That's all. That's enough.

The crew comes back at 6:12 a.m. They're quiet in the way that crews get quiet after something like this — a specific, weighted quiet that is not the same as ordinary exhaustion. They see her in the watch office. One of them — a lieutenant, fifteen years, the kind of officer who does not show much — stops in the doorway for a moment.

'You're here,' he says.

'Yeah,' she says.

He nods once. Goes to change out of his gear. The rest of the crew filters through. She stays until the morning shift comes in.

ChiefBot's documentation is complete by the time she gets back to her desk. Every required notification has been queued. The timeline is accurate. The media statement is ready for her review. The benefit processing has been initiated.

She reads through it. It's done correctly. All of it.

It did not go to Station 7.

Leading Through, Not Around

Some chiefs respond to organizational crisis by accelerating toward normalcy. The LODD happened, the paperwork is filed, the memorial service is scheduled, and by next week the expectation is that the department returns to its standard operating rhythm. This isn't callousness, in most cases. It's a coping mechanism dressed up as leadership — the conviction that forward momentum is the antidote to grief, that keeping people busy protects them from pain.

It's also wrong.

Organizations that are pushed through grief rather than led through it do not recover faster. They develop the organizational equivalent of an untreated wound — scar tissue of suppressed response that surfaces later as attrition, cynicism, interpersonal conflict, and the kind of low-grade institutional dysfunction that no performance dashboard measures and no AI tool detects. The crew that wasn't given space to process what happened doesn't forget it. They carry it, quietly, and it shapes how they work and whether they stay.

Leading through crisis means naming what happened clearly and honestly. Creating formal and informal space for the response. Following up — not once, but repeatedly, over weeks and months, because grief doesn't follow a project timeline. Being willing to sit in a watch office at 6 a.m. with no agenda and no prepared words, because sometimes the leadership act is simply being there when the crew comes home. The International Critical Incident Stress Foundation defines CISM as a multicomponent approach that begins before an incident and extends well beyond it — pre-incident education, acute crisis intervention, and structured follow-up. The IAFF's guidance on CISM identifies the same sequence: minimize the emotional impact, build resistance and resilience, prevent harmful effects through sustained engagement. What both frameworks describe, at their core, is a chief who shows up and keeps showing up. That is not a program. It is a practice.

None of that is in ChiefBot's notification script. None of it can be. The script is the floor — the minimum, necessary, institutional response to an institutional event. What Dana does above the floor is where the leadership lives.

The paperwork was done by the time she got back to her desk. She had been doing something else.

FOR THE READER

This chapter ends with one question, not three.

Think of the most significant crisis your department has faced — an LODD, a serious injury, a mass casualty event, a suicide, a tragedy in the community that hit your people hard. Think about how you responded. Not the administrative response — the paperwork, the notifications, the policy compliance. The human response.

Were you present? Were you present enough? Were you present in the ways that mattered to the people who needed it?

If the honest answer is yes — if you were at that door at 2:30 in the morning, or in that watch office at 6 a.m., or in that station three days later with no agenda — hold onto what made that possible and protect it. The organization you lead is more resilient because of it, even if you cannot measure how.

If the honest answer is no — or not entirely, or not in the ways that mattered most — that's information worth sitting with before you read the next chapter. Not as a verdict on your leadership. As a question about where you want to put your time when AI returns it to you.

ChiefBot will handle the documentation. It will handle the notifications, the timeline, the checklist, the media statement. It will be accurate and complete and timely, and it will free you from the hours of administrative response that have historically fallen on the chief in the immediate aftermath of a crisis.

The question is what you do with those hours.

Dana knew. She drove to the family's home. Then she drove to the station and sat in the watch office until the crew came back.

She was not following a protocol. She was being a chief.

AI will handle the paperwork.

The question is what you do with the time that returns to you.

Chapter 6

The Political Chief

Scene

The city council meeting begins at 7:00 p.m. on a Tuesday.

Municipal Leadership Optimization System, Version 4.2 has been preparing since Monday morning.

The agenda item is a proposed apparatus replacement — a fifteen-year-old aerial platform that has exceeded its manufacturer's recommended service life, accumulated $270,000 in maintenance costs over three years, and been flagged in two consecutive inspections as requiring priority replacement. The cost is $1.2 million. The system has built the case for a supplemental appropriation.

The presentation is, by any measurable standard, exceptional.

A three-year maintenance cost trajectory demonstrating the aerial's accelerating repair burden — $47,000 in year one, $89,000 in year two, $134,000 in the current year to date — with a projected continuation curve that makes the $1.2 million replacement look conservative. A peer department benchmarking analysis showing seven comparable municipalities have replaced aerials of similar age within the past four years, with average replacement costs 12 percent higher than the current quote. An ISO impact analysis. A community risk section identifying fourteen occupancies in the jurisdiction that exceed the reach of the department's remaining ladder truck, with addresses, occupancy types, and estimated life-safety risk scores.

The slide deck is clean. The executive summary fits on one page.

The system presents at 7:14 p.m. The presentation takes eleven minutes. The council asks questions. Each one is answered accurately and completely.

The council votes.

Three in favor. Four opposed.

The item has been logged. Vote outcome: 3-4. The item has been flagged for resubmission in the next budget cycle. A probability model for the next submission has been initiated, pending updated data.

The system moves to the next item in the queue.

It doesn't understand what happened.

Chief Dana Mercer is sitting in the second row when the vote comes in. She watches each Council Member in turn. Hargrove. Vasquez. Park.

She feels something settle in her chest like cold weight.

She understands exactly what happened.

The presentation was flawless. The data was airtight. The case was, by any rational measure, bulletproof.

And she let the data walk into that room alone.

She stays in her seat while the council moves to the next item. She doesn't write anything down. She's spent enough time in that room to know what happened and why and writing it down will not change it. What she is doing, sitting very still in the second row, is running the tape backward. The hallway conversation three weeks ago. The moment she chose not to follow the thread. The briefing request she never knew about until it was too late to matter. Hargrove's vote, which she could not have controlled. Park's vote, which she could have.

She files that distinction carefully. It will matter later.

Story

The vote was never about the aerial.

Council Member Hargrove — the swing vote, the one the system's probability model had assessed as likely favorable based on his public safety voting history — had a conversation with the city manager three days before the meeting. The city manager, facing pressure from the mayor's office about the overall municipal budget, had suggested quietly that this wasn't the right cycle for a supplemental capital request. Hargrove listened. He owed the city manager a favor from a zoning dispute six months ago. He voted no.

Dana had known about the tension between the city manager and the mayor's office. She'd heard it in a hallway conversation three weeks earlier — a clipped exchange that stopped when she walked past, the kind that leaves a residue you cannot quite identify. She had filed it away and kept moving. She hadn't called Hargrove. She had not sat down with the city manager to understand the budget posture before she walked into a supplemental appropriation request. She had assumed that a sound argument would carry the room regardless of what was happening in the rooms she wasn't in.

That assumption was the mistake. Not the data. Her.

Council Member Vasquez, who'd supported every fire department budget request for seven years, was three weeks away from announcing she would not seek re-election. She was tired. She voted the way she thought the mayor wanted her to vote. She had stopped caring about the political consequences of her decisions and started caring about getting through the remaining weeks without conflict.

Council Member Park had asked the fire chief — not the system, the fire chief — for a private briefing on the apparatus situation two weeks before the meeting. She'd sent the request through the city clerk's office, which was her customary channel for formal communications. It was professional, specific, and reasonable: she wanted thirty minutes, she wanted to understand the maintenance cost trajectory, and she wanted to ask questions she didn't feel comfortable asking in a public session.

No one called her back.

Not because anyone decided not to. Because the request arrived during a week when Dana was managing a personnel matter, a mutual aid activation, and the early stages of an accreditation preparation cycle. The clerk's message sat in a deputy's inbox. The deputy assumed someone else had handled it. The briefing never happened.

Park arrived at the council meeting without the conversation she had asked for. She had done her own research. She'd read the maintenance summaries, looked at the peer department comparisons. The data was clear. The case was solid. She knew the aerial needed to be replaced.

She voted no anyway.

Not out of spite. Out of a principle she held quietly and consistently: if a department wanted her support for a significant capital expenditure, they owed her a conversation before they owed her a presentation. The presentation was for the public record. The conversation was for her. One told her what they were asking. The other told her whether they respected her enough to ask properly.

Dana learned about the missed callback three days after the vote, in a passing conversation with the city clerk about an unrelated scheduling matter. The clerk mentioned it almost apologetically — had the chief ever connected with Council Member Park about the apparatus briefing?

Dana felt the specific discomfort of a mistake that was entirely preventable and entirely hers.

She called Park that afternoon. Park took the call. They had the conversation that should have happened two weeks earlier. It lasted forty minutes. By the end of it, Park told Dana directly: bring it back next cycle, call me first, and you'll have my vote.

The aerial replacement was approved seven months later. By a margin of six to one.

The data had not changed. The presentation had not changed.

One relationship had been repaired. That was the difference.

None of what drove those three votes was in the data. None of it could've been. The political landscape that determines whether a budget request succeeds or fails is built from relationships, histories, obligations, grievances, and dynamics that exist entirely outside any dataset the system can access. The presentation was perfect. The case was airtight. And it lost — not because the council disagreed with the analysis, but because three people in that room were operating from contexts the analysis never knew existed.

Dana knew about Hargrove and the city manager. She had felt the shape of it in that hallway conversation and chose not to follow the thread. She trusted the data to do the work that relationships were supposed to do.

That was the mistake. Not the system's. Hers.

Every fire chief operates within a political ecosystem that is, in the most practical sense, the environment in which all of their other work either succeeds or fails. Budget requests, policy changes, mutual aid agreements, staffing decisions, facility investments — every significant initiative the chief pursues requires political navigation. And political navigation is, at its foundation, a relationship skill.

The chief who understands this builds political capital deliberately — the way they build physical fitness, as a practice, not a project. They attend city council meetings when they have nothing on the agenda. They learn what each Council Member cares about and find genuine points of connection to the department's work. They call before they need something, not only when they do.

Marcus Webb arrived at his third department to discover that the previous chief had not attended a city council meeting voluntarily in four years. He had sent the deputy. He communicated through written reports. He had been, by all operational accounts, competent and professional — and he'd accumulated zero political capital in the process. Every budget request started from scratch. Every initiative required twice the effort it should have.

Marcus spent his first six months doing an assessment that had nothing to do with operations. He attended every council meeting, every budget work session, every planning commission hearing. He went to neighborhood association meetings in each of the department's response districts — not to present anything, but to introduce himself and listen. He had coffee with the city manager once a week for three months. He met individually with each Council Member in their offices, with no agenda beyond understanding what they cared about.

In those conversations, he learned things that no dataset contained.

He learned that one Council Member's skepticism about fire department budget requests stemmed from a specific incident six years earlier — a fire department apparatus had damaged a city park and the previous administration had been slow to respond. The incident had been resolved but the skepticism had calcified into a general disposition.

Marcus visited the park on a Thursday afternoon, alone, before he had any specific reason to be there. He stood at the corner where the apparatus had cut across the grass and left ruts that had long since been repaired. He thought about what it meant to walk into a relationship carrying someone else's failure — to be accountable for something you didn't do because the institution you represented had done it. It was an uncomfortable thing to sit with. He sat with it anyway.

In his next conversation with the Council Member, he mentioned the park. Not defensively. Not with an explanation or a reframing. He said: I know what happened there, and I understand why it matters to you. The Council Member looked at him for a moment before responding. That was the beginning of something.

He learned that the mayor's primary concern was not public safety performance metrics but the department's visibility in the community's most economically vulnerable neighborhoods — areas where trust in public institutions was low and where the department had historically been reactive rather than present. Marcus made three visits to community organizations in those neighborhoods in his first ninety days. He didn't bring a presentation. He brought his card and his time.

By the end of his first year, budget conversations were different. Not easy — budget conversations in municipal government are never easy — but different. When Marcus walked into a budget work session, he was walking in as someone the Council Members knew. Some had his cell number. Several called him directly with constituent questions rather than routing through the city manager. One had invited him to speak at a neighborhood event.

In his second year, Marcus brought a staffing request to the council that his predecessor had attempted twice and failed. The data supporting the request was nearly identical to what had been presented before. The request passed four to one.

Afterward, one Council Member who had voted against the identical request two years earlier told Marcus directly: the data was always good. The difference was that this time they trusted the person presenting it.

There is a specific competency embedded in political navigation that deserves its own treatment: reading a room in real time.

A council chamber on a contentious budget night is a dynamic environment. The body language of the swing vote tells you something the agenda does not. The sidebar between the mayor and the city attorney before the meeting tells you something worth knowing. The question a Council Member asks — not just what they ask, but how they ask it, and what they are really asking underneath it — tells you whether the vote is where you thought it was.

The system presented for eleven minutes and received questions it answered accurately and completely. It did not notice that Council Member Park's body language closed when it began speaking — the slight backward lean, the arms that crossed, the attention that dropped from the screen to the notepad in her lap. It did not adjust. It completed the presentation as designed.

A chief who had called Park back two weeks earlier, who had walked into that chamber with a relationship rather than a presentation, would have read the room differently. Not perfectly. Politics is never perfectly legible. But they would have had more information and more options for responding to what they found.

Political navigation doesn't happen only in council chambers. It happens at community meetings, at civic events, in conversations with neighborhood association leaders and school principals and the residents who show up at public hearings to speak for three minutes about something they care about. Every one of those interactions is an opportunity to build or erode the community trust that is, ultimately, the fire department's most durable political asset.

Consider what happens when a department needs to close a station. The operational and financial case is usually straightforward to build — response time modeling, call volume distribution, coverage overlap analysis. The system can produce a defensible analytical case in an afternoon. And in many departments, it will still fail — not because the analysis is wrong, but because the department does not have the community trust to make the conversation possible.

The department that has invested in community presence over years — that has shown up at neighborhood meetings when nothing was at stake, that has built relationships with the civic leaders who shape neighborhood opinion — is not immune to that opposition. But it is in a different conversation. The community leaders who know the chief, who've seen the department's commitment over time, who have been treated as partners rather than managed as stakeholders — those leaders may not support the closure. But they *are* less likely to lead the opposition.

Dana Mercer knows every street in her district. She knows the families who have called 911 more than once and why. She knows the community. But as she sits in the second row watching the vote, she feels for the first time the distance between knowing and tending. Twenty-seven years of accumulated knowledge, and somewhere along the way she had stopped actively investing in the relationships that gave that knowledge its power. She had been present in the operational sense — available when something went wrong, responsive when something was asked of her. She had not been present in the civic sense — showing up without an agenda, maintaining the connections that make the hard conversations possible before they become necessary.

The aerial presentation failed in part because Dana had been managing her political landscape from a distance. She'd been coasting on history when she should have been making deposits.

That's the lesson Council Member Park's vote contained. Not that the data was insufficient. That the relationship was.

Dana doesn't wait for the resubmission cycle to start rebuilding. She schedules coffee with the city manager the week after the vote — not about the aerial, not about anything specific, just to understand where he is and what he's carrying. She attends the next neighborhood association meeting in the district nearest the industrial corridor, something she's been meaning to do for two years. She pulls her calendar for the next three months and identifies every Council Member she hasn't spoken to outside of a formal setting since the start of the fiscal year. The list is longer than she expected.

She isn't doing this because the aerial failed. She's doing it because she understood, sitting in the second row of that council chamber, that she had been managing her political landscape the way she'd been managing her administrative burden — reactively, in pieces, between other things. The system had been doing the data work. No one had been doing the relationship work. That's a gap no algorithm closes.

FOR THE READER

Three questions.

1. Map one key relationship in your current political landscape — a Council Member, a city manager, a union president, a neighborhood leader, a peer chief whose cooperation you depend on — and ask yourself honestly what the system could not know about it. What's the history? What are the unspoken dynamics? What does this person actually care about that doesn't appear in any public record? What would it take to strengthen this relationship before you need something from it? Write it down. Relationships that aren't actively maintained are relationships that are quietly degrading.

2. Identify one relationship in your political landscape that you have been managing from a distance — maintaining through formal communication rather than genuine investment. Not a broken relationship. An under-tended one. What would it cost you to close that distance this month? What might it cost you later if you do not?

3. This month, make one deliberate relationship investment with a political stakeholder — not because you need something from them right now, but because you will. Attend a meeting you don't have to attend. Make a call that's not about a current ask. Learn something about what they are working on and find a genuine point

of connection to your department's work. The best time to build a relationship is before you need it. The second-best time is now.

Dana called Park. She attended the neighborhood meetings. She pulled the calendar and made the list. She wasn't doing any of it because it was comfortable. She was doing it because the aerial needed to come back, and she now understood what needed to come back with it.

The presentation was perfect.

The vote was never about the presentation.

Chapter 7

The Culture Chief

Scene

Station 3 has the highest turnover in the department. Firefighters assigned to Station 3 typically find themselves on one of two paths: requesting reassignment, or finding their way to another agency. The few that find themselves outside these groups are firefighters that never want to work anywhere else in the city.

Municipal Leadership Optimization System, Version 4.2 has known this for eleven months. It has tracked the separations, cross-referenced them against tenure, shift assignment, apparatus type, and geographic posting. It has compared Station 3's attrition rate against the department average and against peer department benchmarks. It has flagged the pattern in three consecutive monthly reports, each time with the same notation: anomalous attrition rate, Station 3, cause undetermined, recommended action: further investigation.

The deputy chief finally asks the system to investigate.

The system recommends a survey.

The survey is well-designed. Thirty-one questions covering job satisfaction, supervisory relationships, workload distribution, equipment adequacy, shift fairness, and organizational communication. Anonymous. Digital. An 84 percent response rate, which the system notes is statistically significant.

The results are analyzed.

Job satisfaction at Station 3 is 23 percent below the department average. Trust in direct supervision is 31 percent below. Perceived fairness of shift assignments is 41 percent below. When asked whether they would recommend the department as a place to work to a qualified candidate, 67 percent of Station 3 respondents said no.

The system generates a twelve-page report. It identifies the statistical patterns. It flags supervisory relationships as the strongest correlate of dissatisfaction. It recommends supervisory training, a review of shift assignment protocols, and a follow-up survey in ninety days.

It cannot tell you who the captain is.

It cannot tell you that Captain Reyes has been at Station 3 for nine years and his leadership style was formed in a department where intimidation passed for accountability. The system cannot tell you that three of the four members who separated in the past eighteen months told their friends — but not HR — exactly why they left. It cannot tell you that the two members who stayed did so because they needed the health insurance and the pension, not because they wanted to be there.

The system didn't tell Dana anything she didn't already know.

It just put a number on it.

Chief Dana Mercer reads the report on a Friday morning. The same desk. The same mug.

She reads the executive summary. Then the supervisory relationship data. Then the attrition numbers broken down by tenure — the members who left in their first two years, the ones who stayed and went quiet.

She sets the report down and looks out the window for a long time.

Two years. She has known for two years. Not from data — from the texture of the place. From a station visit fourteen months ago where the crew was technically fine and personally distant in a way she noticed and couldn't name. From a resignation letter that said more between the lines than in them. From a battalion chief's comment, delivered carefully, that she heard and filed and didn't follow up on.

This is what she doesn't say in any of the reports she reads and any of the reports she writes: she has been managing Station 3 from a distance because going would require a conversation she hasn't been ready to have, and managing from a distance has been easier than admitting that. The data has given her cover. The survey will give her more cover. The follow-up survey recommendation will give her ninety more days of cover after that. The system will generate cover indefinitely, because the system does not know what she knows about why she hasn't gone.

She knows that too.

She has known something was wrong at Station 3 for two years. Not from data — from the texture of the place. A station visit fourteen months ago where the crew was technically fine and personally distant in a way she noticed and couldn't name. A resignation letter that said more between the lines than in them. A battalion chief's comment, delivered carefully, about the captain's style.

She knows. She has known.

She thinks about what she would say if she drove there today. She's composed the conversation in her head so many times she's stopped counting — how she would open it, what she would name, how Reyes would likely respond. She can build the whole structure in her mind: the acknowledgment of the data, the direct statement of what it means, the question that opens the door rather than closes it. She knows what needs to be said. She has known that too. The conversation exists, completely, in her head. It has never existed anywhere else.

What she cannot fully explain, even to herself, is why she hasn't gone.

The report doesn't tell her what to do about that. No report could.

Story

Organizational culture is what happens in your department when you're not watching.

It's the sum of the behaviors that are modeled, tolerated, rewarded, and punished — not in policy, but in practice. It's the answer to the question every new member asks, silently, in their first weeks: how does this place actually work?

The gap between the culture a department espouses and the culture it practices is one of the most reliable predictors of organizational dysfunction, attrition, and — in the fire service — safety failure. A department whose stated values include respect and accountability but whose actual practice tolerates a captain who rules through fear is not a department with a supervisory problem. It's a department with a culture problem. The supervisory problem is a symptom. The culture is the condition.

The system can identify symptoms. It identified Station 3's symptom in eleven months of reports that nobody acted on. What it cannot do is diagnose the condition — cannot go to the station, cannot sit in the dayroom, cannot read the body language when a crew talks about their officer, cannot hear what gets left unsaid.

And it cannot treat the condition, because treatment requires the kind of leadership presence and organizational courage that no algorithm supplies.

The survey the system designed was competent. The analysis was accurate. The recommendations were reasonable. And none of them addressed what was actually happening at Station 3, because what was actually happening at Station 3 was not in the data.

Culture lives in the unofficial spaces. It lives in the conversations that happen after the meeting ends. It lives in who gets the desirable assignments and who does not, and whether the pattern has an explanation anyone is willing to say out loud. It lives in whether a new member's first instinct, when they encounter something wrong, is to speak up or to learn how to live with it.

The survey is not the culture. The data is the shadow the culture casts. The chief who mistakes the shadow for the object is going to design interventions that address the measurement rather than the reality.

Marcus Webb encountered this at his second department, inheriting a culture that looked functional from a distance and felt wrong from the inside. The metrics were adequate. Response times were within standard. The annual climate survey, administered faithfully every October, consistently returned results in the acceptable range.

What the survey didn't capture was that the acceptable range had been achieved through a quiet process of self-selection. Members who found the culture intolerable had left. Members who remained had adapted — had learned the unwritten rules, had calibrated their expectations downward, had made a private accommodation with a workplace that didn't match what they had believed they were joining. The survey measured satisfaction among people who had already decided to stay. It told Marcus nothing about the people who had left, or why.

Marcus spent his first three months at that department in the stations. Not inspecting — visiting. Sitting in day rooms. Eating meals when he was invited. Listening to what people talked about when they weren't performing for a supervisor. He wasn't trying to extract information. He was trying to understand a culture by being present in it, which is the only reliable way to understand a culture.

The moment that crystallized everything came in his eleventh week, in a dayroom at Station 6, in a conversation that started as small talk and became something else entirely.

A six-year member — competent, well-regarded, the kind of firefighter who would have been a natural candidate for promotion — told Marcus, with the careful casualness of someone who'd rehearsed the nonchalance, that he'd decided not to test for engineer. Marcus asked why. The member looked at his coffee for a long moment.

I know what happens to people who test here, he said. I've watched it.

He didn't elaborate. He didn't need to. Marcus had been watching it too. The department had a quiet, reliable pattern: members who pursued promotion were subtly managed out of the informal networks that made daily work bearable. The invitations to shared meals got fewer. The inside jokes landed differently. The belonging that had been unconditional became conditional on not wanting more than your current rank. It was not explicit. It was not policy. It was culture — operating exactly as culture operates, below the level of documentation, through the accumulated signal of a hundred small interactions that collectively said: here, ambition is a betrayal.

That conversation cost Marcus three weeks of sleep. It also gave him the most important piece of information he had gathered in eleven weeks of station visits. Not the satisfaction scores. Not the turnover data. A six-year member who'd stopped wanting to grow because growing felt like losing.

He reorganized the promotional culture within four months. Not the testing criteria — the culture around testing. He named the pattern directly to his command staff, in terms that were uncomfortable and accurate. He remembered the silence in that room when he finished — not hostile, but heavy. The kind of silence that means people are deciding whether to push back or sit with what they have just heard. Two of his deputy chiefs didn't meet his eyes. A battalion chief he had come to trust looked at the table for a long moment before nodding slowly. Nobody said he was wrong. That was not nothing.

He promoted three members in the following cycle who had previously opted out. He made a point of publicly celebrating each promotion in a way that made ambition visible as something the organization valued rather than tolerated.

The six-year member tested in the next cycle. He passed.

He is currently a captain.

Here is the uncomfortable truth this chapter has been building toward: culture does not happen to organizations. It's produced by them, continuously, through the decisions leaders make and the behaviors they model and the things they're willing to tolerate.

Dana Mercer has known something was wrong at Station 3 for two years. That knowledge came from sources the system cannot access — a passing comment from a battalion chief, a resignation letter that said more between the lines than in them, a station visit where the crew was technically fine but personally distant in a way she noticed and didn't follow-up on. She knew. She didn't go.

That's a leadership choice. Not a dramatic one — not a decision made with full awareness of its consequences. The quiet, accumulated choice to manage from a distance rather than lead from proximity. To trust the metrics rather than supplement them with presence. To let a survey substitute for a conversation.

The chief who shapes culture intentionally does several things the system cannot replicate. They are consistent — their behavior in the day room matches their behavior in the command staff meeting. They're visible in the right places at the right times — not the scheduled walk-through that everyone prepares for, but the unannounced visit, the dropped-in meal, the conversation that happens because the chief was there rather than because it was planned. They name what they see — they call out the behavior that contradicts the stated values, not in public humiliation, but in direct conversation that makes clear the behavior is seen and is not acceptable. And they follow up — they don't let a difficult conversation become the last word on a difficult situation.

None of this is complicated. All of it is hard. It's hard because it requires sustained attention in an environment that constantly generates reasons to look away. It's hard because some of the cultural dynamics a chief needs to address involve people with long tenure and the implicit protection of a department that prefers to manage its problems quietly.

Captain Reyes has been at Station 3 for nine years. He is not a villain. He's a product of a culture that made him and a leadership failure that predates Dana. He's never been told, directly and clearly, that what he's doing is not acceptable here. Dana has known this — has sensed it in the texture of that station, in the resignation letters, in the battalion chief's careful comment. She has composed the conversation in her head. She knows what she would say.

She has not gone.

The system flagged the anomaly. It recommended a follow-up survey. It will wait, indefinitely, for a human to decide what to actually do.

Mentorship is one of the most powerful culture-building mechanisms in the fire service, and one of the most inconsistently practiced. When it works — when an experienced member takes genuine interest in a newer member's development, shares not just technical knowledge but institutional wisdom, and creates a relationship that makes the newer member feel that their growth matters to someone — it produces loyalty, retention, and capability that no orientation program generates.

When it doesn't work, the message received is that belonging must be earned through suffering rather than extended through welcome. The newest member watches how the experienced ones treat each other and draws conclusions. The conclusions they draw become the culture they inhabit. And in time, if they stay, the culture they pass on.

Belonging is the felt sense that this organization has a place for you — that your presence is an asset rather than a probationary condition, that the culture has space for who you are and not only who it expects you to be. When belonging is present, it looks like this: a newer member asks a question in front of the crew and nobody makes them feel small for not knowing. A firefighter with a different background from everyone else in the dayroom is included in the conversation rather than tolerated at its edges. A member who struggled through a difficult call finds a senior colleague at their side the next morning, not to debrief but just to check-in.

When belonging is absent, it looks like this: members learn quickly which questions are safe to ask and which reveal too much. People who are different learn to minimize the difference. The ones who cannot minimize it leave. The ones who stay learn to perform a version of themselves that fits — and something is quietly lost in that performance, every shift, accumulating over years into the kind of cultural rigidity that no survey ever fully captures.

In a fire service that's becoming more diverse — in gender, in race, in background, in the life experiences people bring to the job — belonging is not automatic. It's built, deliberately, by leaders who create the conditions for it and who address, directly and consistently, the cultural norms that exclude rather than include.

The system cannot build belonging. It can measure its absence — through attrition data, satisfaction scores, the demographic patterns in who stays and who leaves. Dana is reading that measurement right now, in the Station 3 report. The question is not whether the data is accurate.

The question is what she's going to do about it — and whether she's going to do it in a way that actually changes the culture, or in a way that improves the metrics without touching the reality beneath them.

FOR THE READER

Three questions. Culture requires the same honest engagement as everything else in this book — and more, because culture is the one domain where self-deception is most available and most costly.

1. Conduct an honest culture audit — not a survey, not a dashboard review, but a personal reckoning. Spend one week paying deliberate attention to the gap between your department's espoused values and its practiced ones. Where are the values on the wall not the values in the room? Where is the behavior that contradicts the stated culture being tolerated, and by whom, and for how long?

Write it down. The audit is not for anyone else. It's for you. And it's only useful if you're honest about what you find.

2. Identify one cultural norm in your department that protects the status quo at the expense of performance or belonging. Not a policy — a norm. The unwritten rule that everyone knows, and nobody says. The behavior that's from the top and therefore understood as acceptable throughout. The conversation that never happens because everyone has learned it's not safe to have. Name it specifically. Then ask: what would it cost to challenge it, and what is it currently costing not to?

3. Go to the station you've been driving past. Not to inspect — to be present. Without an agenda. Without a structured outcome. Go because presence is not a policy; it's a choice, and the chief who keeps not making it is the chief who keeps owning the culture that a visit would have changed. Stay long enough to learn something the survey never captured. Then ask yourself: how many weeks has it been since I was somewhere in my department — not virtually, not in a meeting, but physically in a place my people work — without a reason to be there except that I wanted to understand what was happening? The answer to that question is the measure of the distance you have allowed to accumulate. Culture doesn't close that distance. Only going does.

The system knows the numbers.

It doesn't know the station. Only you can go.

Chapter 8

The Ethical Chief

Scene

The recommendation arrives on a Monday morning, formatted as a memo.

Municipal Leadership Optimization System, Version 4.2 has completed its analysis of the department's station coverage optimization study — a project initiated eight months earlier in response to a city council request for a long-range facilities assessment. The study has been thorough. The system has analyzed response time data across the department's six stations, call volume distribution by geographic zone, apparatus utilization rates, population density projections through 2035, infrastructure costs, and capital maintenance schedules for each facility.

The findings are clear.

Station 4 is the least efficient station in the department. It serves the lowest call volume per apparatus hour of any posting. Its infrastructure requires $340,000 in deferred maintenance over the next five years. Its response area overlaps significantly with Stations 2 and 6, which between them can cover 91 percent of Station 4's current response zone within acceptable NFPA response time benchmarks. Closing Station 4 and redistributing its apparatus and personnel to Stations 2 and 6 would reduce annual operating costs by $1.2 million, eliminate the deferred maintenance liability, and improve overall apparatus utilization across the department.

The recommendation is: close Station 4.

The system notes two items flagged for human review: community impact assessment, which it has outlined but noted requires local context it cannot supply; and equity implications, which it has identified as a relevant consideration but assessed as outside the scope of its analytical framework.

It queues the memo for Chief Mercer's review and moves to the next task.

It doesn't know where Station 4 is.

It doesn't know that Station 4 sits at the corner of Delmar and Fifth, in the oldest neighborhood in the city, in a district where the median household income is $31,000 and the district is majority-minority. It doesn't know that the station has been in that building for sixty-one years, that three generations of families from that neighborhood have watched apparatus roll out of those doors, that the department's relationship with that community was rebuilt — slowly, painfully, over a decade — after a period of institutional neglect that the neighborhood has not forgotten and will not soon forgive.

It doesn't know that the second-closest alternative coverage — Station 6 — is eleven minutes away on a bad traffic day. And in that neighborhood, there are a lot of bad traffic days.

Dana Mercer reads the memo at her desk. Not the kitchen counter this time. She is already at work, already into her second hour of the day, when the notification arrives.

She reads it once. The numbers are familiar — she has reviewed the underlying data herself over the past eight months. The system has not made an error.

She reads it again.

There is a specific sensation she has not felt before in all the months of reading ChiefBot's output — not the vertigo of Chapter 1, not the quiet anger of Chapter 2, not the uncomfortable recognition of seeing her own name in the training profile. This is something heavier. The feeling of holding two true things simultaneously and knowing that only one of them can govern.

Station 4 is, by every operational metric, the most rational candidate for closure.

She thinks about the faces at the community meetings when the facilities study was announced — the wariness that settled over the room before anyone had said a word, the specific patience of people who have learned not to assume good faith from institutions. She thinks about the retired firefighter who grew up on Delmar, who volunteers at the station on weekends, who told her once that when he was a kid that building was the only city-owned property in the neighborhood that didn't feel like a place designed to process people. She thinks about what it took — what it actually took — to rebuild the department's relationship with that community after the years it was allowed to deteriorate.

She closes the memo.

The numbers are right. The answer is wrong.

Story

There's a version of what just happened that ends with Dana implementing the recommendation.

It's not because of corruption, laziness, or being short on time. It's the version in which a chief under budget pressure, facing a city council that asked for efficiencies, accepts an analytically sound recommendation from a system that has been right about everything else. The version in which the data carries the decision because carrying the decision is what data is for.

It's the version in which nobody does anything wrong. And Station 4 closes. And the neighborhood absorbs it the way neighborhoods absorb these things — with anger that eventually becomes resignation, with a loss of trust that does not announce itself but calcifies quietly, with a relationship that took ten years to rebuild and will not survive the decision.

That is the version where the algorithm wins by default. Not because it was right. Because nobody checked its work.

Data-driven decision making is one of the most important shifts the fire service has made in the past two decades. It has produced better resource allocation, more defensible budget requests, more precisely targeted prevention programs, and operational improvements that genuine analysis revealed, and instinct alone never would've found. The argument for using data — made throughout this book — is real and is not retracted here.

But data is not values. And values are not data.

Efficiency and equity are not always opposites. In many decisions they point in the same direction — the well-targeted prevention program that reaches the highest-risk population is both more equitable and more effective than the one that serves the path of least resistance. But when they conflict — when the efficient choice and the right choice are not the same choice — the chief has to decide which one wins. That decision cannot be made by a system. It can only be made by a leader who has thought carefully about what the department is actually for.

The system's recommendation was analytically correct by every metric it was configured to measure. What it was not configured to measure — what no system can be configured to measure — is what a community has been through, what a relationship with a public institution has cost a neighborhood, and what it means for a fire department to make a decision that the community it serves will experience as abandonment.

Those things are not irrational. They are not soft. They are the accumulated history of people who have been making calculations about whether to trust public institutions for longer than any single chief officer has been doing this job. They belong in the decision. They are not in the data.

The fire chief who uses AI to inform decisions without using judgment to govern them is not leading. They are delegating leadership to a system that was never designed to carry it.

The department's relationship with the Delmar and Fifth neighborhood wasn't rebuilt through a policy change. It was rebuilt through a decade of deliberate, unglamorous work — a chief who showed up at neighborhood association meetings without an agenda, who made sure the station's community room was available for meetings and tutoring programs and whatever the block association needed that week, who assigned a community liaison who actually lived in the district and knew its people by name. It was rebuilt through a smoke detector installation program that prioritized the neighborhood's aging housing stock, through a Citizens Fire Academy that recruited participants from Delmar specifically, through the slow accumulation of small interactions that, over time, changed what the station meant to the people who lived around it.

None of that appears in the utilization metrics. None of it registers in the call volume data. The system sees a station with low efficiency numbers. It doesn't see sixty-one years of institutional history, ten years of deliberate repair, or the specific fragility of trust that's been broken once and knows it can be broken again.

Ethical judgment in fire service leadership is not primarily about dramatic choices, and rarely about line-of-duty deaths, catastrophic failures and the kinds of decisions that get written about afterward. It's mostly the accumulation of smaller choices — who gets resources, which communities receive prevention programs, and which don't, which neighborhoods see the department as a partner, and which see it as a presence that arrives when called and otherwise remains invisible.

These smaller choices compound. They produce the culture of a department's relationship with its community, and they produce the conditions under which the larger choices either go well or go badly.

The equity dimension of fire service decision-making is not a separate consideration from operational excellence. It's a component of it. A department that deploys prevention resources based on where its relationships are strongest rather than where the data says risk is highest is not serving its community well — it's serving its comfort zone. A department that makes facilities decisions based on utilization metrics without accounting for the historical context of those metrics — for the reasons why some neighborhoods have lower call volumes, for the ways in which decades of underservice shape the patterns the data now reflects — is not making neutral decisions. It's laundering past choices through present mathematics.

Marcus Webb understood this at his third department, where he inherited a prevention program that had not been updated in eight years and a community risk analysis that showed, on its face, that the department's highest-risk zone was the commercial district downtown. The data was accurate. It reflected what the records showed.

What the records did not show was that the residential neighborhoods in the department's southern sector had been systematically underserved — not out of malice, but out of the accumulated inertia of going where the relationships already were. The lower incident rate in those neighborhoods was not evidence that they were safer. It was evidence that the department had not been paying attention to what was happening there.

When Marcus reoriented the prevention program toward where the risk actually was, the internal resistance was immediate and genuine. His fire prevention captain — a fifteen-year veteran who had built the commercial district program from nothing and was proud of what it had become — sat across from Marcus and said, quietly but directly: we're abandoning what works.

Marcus understood what the captain meant. The commercial district program worked. The relationships were strong, the compliance rates were good, the outcomes were measurable. What Marcus had to name — and did name, in that conversation and in the command staff meeting that followed — was that working and sufficient were not the same thing. That the southern sector neighborhoods had not been experiencing fewer fires because the department had done good prevention work there. They had been experiencing different patterns of harm, less visible harm, harm that the existing metrics were not designed to see.

The captain did not agree immediately. He came around over the following months, as the data from the reoriented program began to surface what had been invisible. But Marcus remembered the discomfort of that transition — the specific cost of naming a truth that implicated people who hadn't been trying to fail anyone, who'd simply been operating within the inherited logic of a program they hadn't designed and had not thought to question.

That discomfort is part of the job. The chief who leads only the comfortable version of ethical decision-making is not leading ethically. They're managing appearances.

The system flagged equity implications as outside the scope of its analytical framework. That notation is the most honest thing the system has ever produced.

Outside the scope of its analytical framework is exactly where the most important decisions live.

A system that recommends closing Station 4 has done its job. It has analyzed the operational data, identified the efficiency opportunity, surfaced the recommendation, and flagged what it cannot assess. It left the rest to the human.

The human's job is not to validate the recommendation or reject it. It's to ask the question the system cannot ask: what are we actually deciding here, and who bears the cost?

Closing Station 4 saves $1.2 million. It also tells a community with a long memory that the fire department's commitment to Delmar and Fifth was contingent on efficiency. That the relationship rebuilt over a decade of deliberate investment can be undone by a spreadsheet. That when the numbers say go, the department goes — regardless of what going means to the people left behind.

Dana has to decide whether that's the department she's running. No system can decide that for her.

What a chief owes their community is not efficiency. Efficiency is a means. What they owe is honest stewardship — of resources, of relationships, of the public trust that is the fire service's most fundamental asset and its most fragile one. The chief who treats that trust as a given, who assumes it survives decisions made without reference to the people affected by them, is working from a model of leadership that the fire service has paid for before and isn't finished paying for now.

She learns two days later, in a passing exchange with her deputy, that the city manager had already seen the recommendation. ChiefBot's summary had been distributed automatically through the facilities study integration — one of the Version 4.2 outputs she hadn't known was going directly to his office. He had indicated, informally and through channels she had not been part of, that the efficiency case was sound. The deputy mentioned it the way deputies mention things they assume you already know.

Dana had not known. The decision had moved without her. Not maliciously — efficiently. She filed it and kept working.

Dana closes the memo.

She opens her calendar.

She schedules a community meeting at the Delmar and Fifth neighborhood center for the following Thursday.

She doesn't know yet what she will say. She knows those rooms — the specific quality of attention in a community that has been through this before, that knows how to listen to the opening sentences of a presentation and determine, in the first two minutes, whether what's happening is a conversation or a performance. She has been in rooms like that. She knows she has to earn the right to speak in that room in a way that she doesn't have to earn it anywhere else.

She knows she owes them the conversation before she owes the city council the answer.

FOR THE READER

Three questions. Ethical judgment requires the same honesty as every other challenge in this book — and the willingness to examine not just what you decided, but how you decided it.

1. Identify one significant decision your department made in the last two years that was driven primarily by data — a resource allocation, a staffing model, a prevention program focus, a facilities choice. Now ask: was there a values check? Did anyone in the room ask who bears the cost of this decision, and whether that distribution is one the department can stand behind? If the answer is no — or if you are not sure — that's worth examining. Data without a values check is not neutral. It's a choice made without full accountability.

2. Name one equity gap in your community risk profile that your department hasn't addressed — a neighborhood underserved relative to its risk, a population with lower engagement in your prevention programs, a part of your response area where the relationship between the community and the department is thinner than it should be. Not a gap you were unaware of. One you know about. Ask yourself honestly: why hasn't it been addressed? The answer is usually not resources. It's usually priority — which is to say, values.

3. The next time a data-driven recommendation lands on your desk — a resource reallocation, a facilities decision, a prevention program adjustment — before you accept it, ask one question the system did not ask: who bears the cost of this if we're wrong? Not the department. The community. The neighborhood. The population that the recommendation most directly affects. The system's job is to identify the efficient option. Your job is to decide whether efficiency is the right governing value for this particular decision. Sometimes

it is. Sometimes what looks like efficiency is the arithmetic of accumulated neglect, accumulated judgement. The chief who can tell the difference is the one doing the job the system cannot do.

The system's recommendation was analytically correct.

The chief's job is to decide what correct means.

BRIDGE

What You Actually Bring

Bridge

What You Actually Bring

Scene

The request comes from Dana herself.

It is the first time she has asked the system for something that was not operational, administrative, or analytical. It is a Tuesday evening, late, and she is alone in her office with the door closed and the building quiet around her. She has been sitting with the Station 4 memo for three days. She has scheduled the community meeting. She has not yet told the city manager what she intends to recommend.

She opens the system interface and types a prompt she has been turning over in her mind since Monday morning.

List everything you cannot do.

The system processes the request for 1.7 seconds — longer than usual, which means nothing, because the system does not experience duration. It is simply a larger query.

The list arrives.

It is four pages long.

She reads it slowly, the way she used to read incident after-action reports when she was a battalion chief — looking for the pattern underneath the individual entries, the thing that connected the specific failures into a larger shape.

Cannot deliver a notification to a family with the presence the moment requires.

She thinks of the Okafor family's porch light. The way Amara Okafor opened the door before Dana knocked, as if she already knew.

Cannot sit with a family at 2:30 in the morning.

She had stayed for two hours. Not because policy required it. Because leaving felt wrong.

Cannot read a room in real time.

She thinks of Council Member Park's arms crossing as the presentation began. The slight lean back. The attention that dropped to the notepad. She had seen it from the second row and done nothing with what she saw, because she had trusted the data to compensate for the relationship.

Cannot identify the cultural dynamic beneath a statistical pattern.

Station 3. The survey score. The thing the survey could measure and the thing it couldn't. She has known something was wrong there for two years.

Cannot assess community trust or its fragility.

Cannot make a values-based decision when values and data conflict.

Delmar and Fifth. The $1.2 million. The sixty-one years. The numbers are right; the answer is wrong. She had known that before ChiefBot told her.

Cannot build a relationship and follow up.

Cannot read the wariness in a room full of people who have learned not to assume good faith from institutions.

She thinks of the Delmar and Fifth community meeting. The one she has scheduled for Thursday. The one she has been dreading because she knows what those rooms are like when the department has not earned the benefit of the doubt and has to work for it in real time.

Cannot have the conversation the situation requires.

Reyes. Captain at Station 3. She has composed that conversation in her head forty times in two years. She has not had it.

Cannot be present.

Cannot go.

She sets the document down.

She has been the one who could go. For eleven years, she has been the one who could go. The one the department sent when presence was what was needed. The one who drove to the Okafor family's home at 2:30 in the morning because a notification script was not enough. The one who knew every street and every alderman and every shortcut in this district because she had been in this district, in this building, doing this job for longer than most of her members have been alive.

She has been the one who could go.

And she has not been going.

Not to Station 3. Not to the community meetings in the Delmar and Fifth district, where she has been sending a deputy for four months because the deputy is capable and the meetings are on Tuesday evenings and Tuesday evenings have been consumed by other things. Not to the conversation with Reyes that she knows needs to happen and has been managing around rather than through. Not to the hard versions of the things she has been doing in their easier approximations.

The system did not say that. The list did not say that. The list says only what the system cannot do. What she has been failing to do is the implication she has been avoiding since the first morning — the question underneath the question, the one she has been successfully not answering for eight months.

Reading the list, she cannot avoid it anymore.

The system's limitations map onto her own avoidances with a precision that is almost comic in its specificity. Cannot be present. Cannot go. Cannot have the conversation the situation requires. She has been telling herself those are the system's gaps. They are. They are also hers.

That is what shifts.

Not the discovery that AI has limits. The recognition that its limits and her limits are not the same limits. The system cannot go to Station 3. She can. The system cannot sit with a family at 2:30 in the morning. She did, once, and it was the right thing and she has been treating it as an exceptional act rather than the ordinary baseline of what her job requires.

She closes the interface. Sits in the quiet for a while.

The building hums around her the way buildings hum late at night — the HVAC, the refrigerator in the break room, the specific ambient presence of a structure that exists for a purpose. Somewhere across the city, her department is operating. Crews are sleeping in stations that are not sleeping, waiting in the particular suspended readiness she knows in her body from years of living inside it.

She picks up a pen. Pulls a notepad from the desk drawer.

She writes: Station 3. Reyes. Thursday.

Then: Delmar and Fifth. Community meeting. Go myself.

Then she sits for a moment and writes one more thing, not a task but a question she has been not asking for eight months, the one that has been clarifying itself in the white space of ChiefBot's limitations list:

What would I be doing with my time if I were doing the job the system cannot do?

She does not have the full answer yet. But she has the question, which is where every real answer starts.

Story

Neither what the system can do nor what it cannot is the point.

The point is what you do next.

Dana Mercer has spent eight months watching Municipal Leadership Optimization System, Version 4.2 work. She has watched it complete in four hours what consumed her last three weeks. She has watched it find a pattern in five years of incident data that nobody in the department had seen. She has watched it build a training needs assessment that revealed something true and uncomfortable about her own performance. She has watched it draft a strategic plan and win a budget presentation it should have won and recommend a station closure that was operationally correct and humanly wrong.

She has been unsettled by it, frustrated by it, quietly threatened by it, and — in the case of the Station 4 memo — genuinely grateful for the flag it left in the margin that said: this requires judgment I do not have.

What she has not done, until tonight, is ask the system what it cannot do. She has been so occupied with what it can do — so caught in the vertigo of watching her job reflected back to her in four-hour reports and 94-page plans — that she forgot to look at the other side of the ledger.

The list is not a surprise. She knew, abstractly, that the system had limits. She has known it since driving to the Okafor family's home at 2:30 in the morning and understanding in her body what a notification script could not contain. She has known it since watching the system present flawlessly to a council that voted no because the relationships were wrong. She has known it since reading the Station 3 survey and recognizing that the number the system put on her inaction was not new information — it was information she already had, had been carrying for two years, and had chosen not to act on.

The list crystallizes what she has known. It maps the system's limitations onto the exact terrain of her own avoidances.

That is what shifts.

Not the discovery that AI has limits. The recognition that her limits and the system's limits are not the same limits. The system cannot go to Station 3. It cannot sit with a family at 2:30 in the morning. It cannot read the wariness in a room full of people who have learned not to assume good faith from institutions. It cannot have the conversation with Reyes that the situation requires.

She can.

She has been able to all along. The capacity was never the question. The choice was.

This is the moment the book has been building toward — not a revelation, but an accumulation. Eight months of the system doing things Dana expected to be hers, revealing things she expected to remain invisible, flagging things she expected to manage from a distance. Eight months of a system being right about what it could measure and silent about everything it could not, and Dana filling that silence with avoidance rather than action.

The list does not tell her what to do. It tells her what she is for.

She is for the things the system cannot do. Not all of them — some of what the system cannot do is simply beyond any single human too. But the ones that are hers. The presence. The relationship. The conversation. The willingness to go to the places the data can point to but cannot enter.

The chiefs who will lead most effectively in the next decade are not the ones who resist AI or the ones who defer to it. They are the ones who understand, specifically and practically, where the system ends and where they begin — and who invest their time, their attention, and their presence in exactly that territory.

That investment does not happen automatically. It requires the same deliberateness that the system applies to everything it does — the same specificity, the same follow-through, the same refusal to let the uncomfortable item sit in the queue indefinitely.

Dana has a notepad on her desk with three things written on it. Station 3. Reyes. Thursday. Delmar and Fifth. Community meeting. Go myself. And one question she has been not asking for eight months.

She does not know yet whether she will do all three. She has written things down before and found reasons not to act on them. What is different tonight is that she understands, for the first time with full clarity, what the cost of not acting has been — not to her career or her reputation or her standing with the city manager, but to the places and the people that needed her to go and found, when they looked for her, a deputy or a report or a well-formatted survey result.

That is what changes in Part Three.

FOR THE READER

One question. This is a bridge, not a destination.

Before you go further — before you engage with the practical question of how to work alongside AI, how to build the habits that keep you relevant, how to lead your organization through the change that is already underway — do what Dana did.

Open something. A notepad. A blank document. Whatever you use when you are thinking rather than reporting.

Write down the things that are on your list. Not the system's list. Yours. The stations you have been driving past. The conversations you have been composing in your head without having. The relationships you have been managing from a distance because managing from a distance is easier than the alternative. The places that have been waiting for you to go and have been getting a deputy, a report, a well-formatted result instead.

The system cannot do those things.

You can.

The question is whether you will.

GO.

PART THREE

The Augmented Chief

Chapter 9

Working With the Machine

Scene

It starts with the schedule.

Dana has been thinking about Marcus's story from the first chapter — the color-coded spreadsheet, the twelve hours a month returned to a chief officer who hadn't known he was losing them. She has been thinking about it on and off for weeks, the way you think about something you know you should do and keep finding reasons not to.

On a Wednesday morning, three days after she made her list, she does it.

She spends ninety minutes configuring a scheduling tool. She imports the department's leave request data, the minimum staffing requirements, the holiday rotation, the apparatus certification constraints. She runs it. It produces next quarter's schedule in four minutes, identifies two conflicts she would have caught eventually and one she would not have, and formats the output in the department's standard template.

She sits back and looks at it.

It is correct. It is better than what she would have produced in three afternoons of manual work, because it caught the conflict she would have missed, and it did not get tired halfway through.

She saves the file. She marks the task complete. She looks at her calendar and realizes she has just returned to herself approximately six hours over the next three months that she was going to spend building that schedule by hand.

She does not know what to do with six hours.

She checks her calendar while the scheduling tool is still open on her screen. The monthly briefing with the city manager — first Thursday, forty-five minutes, eleven years of first Thursdays — has been rescheduled indefinitely. The message from his assistant arrived two days ago. Dana had read it and kept moving. She tells herself it is his calendar. He is a busy person. She believes it, mostly.

She closes the calendar. Opens a new integration tab. Keeps working.

That is the problem she has been avoiding without naming it. Not the tools. Not the learning curve. Not the time it takes to configure something new. The deeper problem: she has spent eleven years filling every available hour with the work in front of her, and she does not have a practiced answer to the question of what she would do if the work in front of her took less time.

The system has been asking her that question for eight months.

She is finally ready to answer it.

Story

The practical question of how to work alongside AI is simpler than most chiefs expect and more specific than most AI advocates admit.

It is simpler because the tools are more accessible, more affordable, and more immediately useful than the word artificial intelligence tends to suggest. It is more specific because the tools that work well for scheduling are not the same tools that work well for grant writing, and the approach that produces useful output for policy drafting is not the same approach that produces useful output for incident analysis. Working with AI effectively is a learnable skill. It is not a single skill. It is a collection of them, each matched to a domain.

Dana's scheduling experiment illustrates the core principle: AI works best on tasks that are rules-based, data-rich, recurring, and time-consuming. The schedule meets all four criteria. So does compliance tracking, policy review, budget narrative drafting, grant opportunity identification, and training needs assessment — the administrative functions that have been consuming the better part of Dana's professional week for eleven years.

The question is not whether AI can help with these things. It can. The question is how to configure that help so that it actually produces time rather than simply relocating the work.

Prompt literacy is the foundational skill.

Every AI tool — whether a scheduling platform, a large language model, a records management analytics suite, or a grant research tool — responds to input. The quality of that input determines the quality of the output. This is not a technology insight. It is a communication insight. Vague requests produce vague results. Specific, well-structured requests with clear context and explicit parameters produce results that are actually usable.

The difference is visible in a direct comparison. Consider two prompts for the same task:

> *Vague: "Draft a policy on apparatus response protocols."*
>
> *Specific: "Draft a department policy on apparatus response protocols for a combination department with four career stations and two volunteer stations, consistent with NFPA 1750 standards, addressing minimum staffing requirements, response time objectives, and apparatus dispatch criteria. Use numbered sections and action-oriented language consistent with our existing policy manual format. Flag any areas where career and volunteer protocols may differ and require separate treatment."*

The first prompt produces something generic that requires significant reconstruction before it is usable. The second produces a working draft that requires editing — the kind of editing a chief can do in twenty minutes rather than the kind of rebuilding that takes an afternoon. The tool is identical. The prompt is the difference.

Learning to write effective prompts — to give AI tools the context, constraints, and output specifications they need to be useful — is the single most transferable skill in working with AI. It takes practice. It is not difficult. And it compounds over time, because every prompt you refine becomes a template you can adapt.

Marcus Webb understood this early. At his third department, he spent two hours one Saturday morning writing what he called a department context document — a one-page description of the department's size, structure, staffing model, response area, accreditation status, and the specific formatting conventions of its policy manual. Every time he used an AI tool for department work, he pasted that document at the beginning of the conversation. The output improved immediately and continued to improve as he refined the document over time. What had taken an afternoon of back-and-forth to produce usable results began producing usable results in twenty minutes.

Two hours on a Saturday morning. That investment returned itself within a week.

Dana builds hers on a Saturday morning three weeks into the experiment. She opens a blank document and types for forty minutes. Department size. Staffing model. Response area. Accreditation status. Union contract provisions that affect policy language. The specific heading format the department has used since the early 2000s that nobody has ever written down but everyone who produces policy documents instinctively maintains. She writes down the things she knows about her department that exist only in her head and have never existed anywhere else. By the time she finishes, the document is one page and change. She reads it back and thinks: this is what I carry. This is what eleven years looks like when you try to write it down.

She pastes it at the beginning of every AI conversation she has about department work from that point on. The difference in output quality is immediate and significant.

Build a personal AI workflow, not an AI experiment.
The chiefs who get the most value from AI tools are not the ones who try the most tools. They are the ones who identify three to five high-value, recurring tasks and configure reliable AI support for each of them — then use that support consistently rather than sporadically.

The experimental approach — trying a new tool when something interesting appears, using it once or twice, setting it aside when it produces an imperfect result — generates insight without generating return. The chief who has tried twelve tools and integrated none of them is not ahead of the chief who has tried two tools and built both into their weekly workflow.

The workflow approach starts with a simple audit: where does your time actually go? Not where you intend it to go — where it goes. Most chiefs who do this exercise honestly are surprised by the results. The gap between intended and actual time allocation is usually significant, and it almost always reveals more administrative consumption than the chief realized.

For most fire chiefs, five integrations cover the majority of administrative drift that consumes a professional week: scheduling and staffing, where AI tools process constraints and produce optimized outputs faster and more accurately than manual methods; training compliance tracking, where continuous monitoring and proactive flagging prevent last-minute scrambles; policy review and drafting, where AI tools flag outdated documents, draft updated language, and queue items for human review; report drafting, where recurring reports can be templated and partially automated so the chief's role becomes review rather than composition; and grant research, where AI tools monitor opportunity databases and match department characteristics against eligibility criteria. These five together are a reasonable six-month workflow build — not a lifetime project, not an overnight transformation. One integration per month, configured properly, used consistently.

Dana starts with scheduling because she already has. By the end of month one she adds training compliance tracking — she has seventeen members whose certifications lapse in the next six months and she has been managing the calendar manually. The tool flags them automatically, drafts notification emails for her review, and updates the tracking record without requiring her to touch it. She reviews the flagged items weekly. It takes twelve minutes. The same work used to take two hours on the first Monday of every month, and she was never fully confident she had caught everything.

By month three she has the policy review integration running. By month five, grant research. Each integration builds on the one before it — the context document seeds all of them, and as she refines it the outputs across every tool improve simultaneously. What had felt, in August, like an insurmountable learning curve looks, by February, like a Tuesday morning.

Protect what you reclaim.
This is the step Dana almost skipped. She returned six hours to herself by configuring the scheduling tool, and her immediate instinct was to fill those hours with other administrative work — the compliance report she was behind on, the policy review she had been deferring, the meeting she had been too busy to schedule.

She caught herself.

The point of reclaiming time from administrative work is not to create capacity for more administrative work. It is to create capacity for the work that only the chief can do — the work the system flagged on its limitations list. The relationship that needs tending. The conversation that needs having. The station that needs a visit.

Dana blocked three of the six hours on her calendar. Not for meetings. Not for reports. She labeled them leadership work and she protected them the way she would protect an operational commitment. She did not fill them immediately. She sat with the unfamiliarity of having unscheduled time and let herself get used to it.

The first week she spent one of those hours visiting Station 3.

She did not have a plan. She did not have an agenda. She drove there on a Tuesday afternoon and walked into the dayroom and said she was just stopping by.

She stayed for two hours.

She did not talk to Reyes that day. But she was there — present in the specific way that cannot be scheduled or optimized or delegated. She sat with the crew at the kitchen table and drank bad coffee and listened. She learned that the newest member, a firefighter eighteen months on the job, had been driving twenty-six miles each way because she couldn't afford to live closer to the station on her starting salary. She learned that the apparatus bay had a slow hydraulic leak that had been on the maintenance list for six weeks and hadn't been looked at. She learned that the crew called the dayroom television the hostage because the remote had been missing for three months and it was stuck on the same channel.

Small things. Human things. Things that were not on any report and would not appear on any survey and could not be flagged by any system.

She drove back to headquarters with the window down, thinking about the firefighter's commute and the hydraulic leak and the missing remote, and she felt something she had not felt in a long time — not the vertigo of the past eight months, not the accumulated weight of avoidance, but the specific satisfaction of having been exactly where she was supposed to be.

That is what the tool returned to her. Not six hours. The possibility of those two hours at Station 3.

FOR THE READER

Two questions. Both require action, not just reflection.

1. Audit where your time actually goes this week — not where you intend it to go. Track your time in thirty-minute increments for five working days. At the end of the week, categorize every block: tasks that legally or practically require your personal judgment; tasks that require your approval but not your labor; tasks being done by you simply because you are the one doing them. The third category is

your AI workflow starting point. Identify the three highest-volume items in that category. Those are the tasks you configure first.

2. This month, build a personal AI workflow document — one page, no more. Describe your department: size, staffing model, response area, accreditation status, apparatus complement, union status, formatting conventions for your policies and reports. This is your context document. Paste it at the beginning of every AI tool conversation you have about department work. Refine it as you learn what produces better results. This single document will improve the quality of every AI interaction you have for as long as you use it — and it will take you two hours to build.

Then ask yourself what you will do with the time those two investments return.

Not what you should do. What you will actually do.

The tool returned six hours.

She spent two of them at Station 3.

Chapter 10

Staying Relevant

Scene

Municipal Leadership Optimization System, Version 5.0 arrives on a Thursday morning in the form of an email from the city's IT department. The subject line reads: Scheduled System Enhancement — No Action Required. The body is three paragraphs of cheerful technical language describing expanded capabilities, improved integration, and what the email calls 'enhanced stakeholder transparency features.'

Dana reads it twice. The phrase that stops her isn't artificial intelligence or autonomous processing. It's stakeholder *transparency.*

She looks it up in the release notes attached to the email.

Stakeholder Transparency Features: Version 5.0 includes automated distribution of key performance indicators to designated city leadership recipients. Reports are generated and distributed on a configured schedule without requiring department-level prompting. Designated recipients may access real-time dashboard data at any time through the city's unified administrative portal.

Dana sets her phone down on her desk. The same desk. The same mug.

She picks it up again and reads it one more time.

Then she opens her calendar and looks at the monthly briefing she has with the city manager. She has been doing this briefing for eleven years — the first Thursday of every month, forty-five minutes, her summary of where the department is and what it needs. She knows which items he cares about. She knows how he takes his coffee. She knows that he reads the budget summaries but that what he actually trusts is what she tells him in that room, in her voice, with her context.

The briefing is on the calendar.

For now.

The first call comes eight days later. Not from the city manager. From his deputy.

The call is professional and brief. The deputy is confirming a figure from the Version 5.0 apparatus utilization report — a discrepancy between the automated summary and a number in a supplemental document. Dana explains it. The deputy thanks her. The call ends in four minutes.

Dana holds the phone after the call ends. She has been doing this job for eleven years. In eleven years, she has never been called by a deputy to verify a figure that she wasn't also in the room to explain. She has been the one who explains. She has been the one who walks into the room.

She files it. Keeps working.

The second sign comes two weeks after that.

Council Member Park calls — not for the quarterly briefing, not about a specific agenda item, but to confirm something. She had read a Version 5.0 community risk summary that flagged elevated heat-related call volume in the Delmar and Fifth district during the previous two summers. She wanted to know if the department had a prevention response.

Dana knew about the call volume. She'd been planning to bring it to the community advisory meeting the following month. She had context that the summary didn't have — the specific buildings, the population, the relationship with the block association that made a targeted outreach program feasible in a way that a generic prevention push was not.

She told Park all of it. Park listened. She thanked Dana and said she appreciated the context.

After she hung up, Dana sat at her desk for a long moment. Park had called to confirm. Not to ask — to confirm. The information had already arrived. Dana had been invited to add texture to what the system had already said.

That is not the same thing as being in the room.

The third sign is the briefing.

It disappears from the calendar on a Wednesday afternoon. The message from the city manager's assistant is three sentences: his schedule has shifted, the briefing will be rescheduled when availability aligns, she'll be in touch. The tone is routine. Entirely routine.

Dana reads it twice.

She had seen a similar message three months earlier, when the monthly briefing disappeared while she was building her AI workflow. That time she'd told herself it was his calendar. The city manager was a busy person. She believed it.

She doesn't call to ask. Instead, she opens her laptop and navigates to the city's administrative portal — the one she now has access to as part of the Version 5.0 integration. Seven dashboards. Response time trends. Budget variance. Apparatus utilization. Community risk indicators. Training compliance. Staffing levels. Incident volume by district.

Everything she has been bringing to the monthly briefing for eleven years is there. Updated automatically. Available at any hour. No appointment required.

She closes the portal. She does not close her laptop. She sits at her desk for a long time, looking at the screensaver, which is a photograph of the department she has led for eleven years.

She's not sure, for the first time, that it will always be hers to lead.

What happens next is small. She'll think about it for a long time afterward.

She's in city hall three days later for a budget pre-meeting with the finance director. She arrives four minutes early — she always arrives early; it's an eleven-year habit — and she's standing in the corridor outside the conference room when she hears the city manager's voice through the partially open door of his office.

He's not talking to her. He's talking to someone she cannot see. His tone is satisfied, the specific satisfaction of a person who has found a more efficient way to do something.

' — which is why I stopped scheduling the department briefings. The portal gives me everything I need by seven a.m. The trends, the flags, the utilization numbers. I can pull it myself in ten minutes instead of sitting in a forty-five-minute meeting.'

A pause. The other voice says something she can't hear.

'Dana's good,' the city manager says. 'The department's performing. I'm not saying that. I'm just saying I don't need the briefing anymore. The system tells me what I need to know.'

The finance director appears at the end of the corridor. Dana turns. She smiles. She follows him into the conference room.

She doesn't remember anything that was said in the budget pre-meeting. She remembers sitting in her car in the city hall parking garage afterward, hands on the steering wheel, engine off.

Eleven years.

She has been a fire chief for eleven years in this city. She has sat with this council through three budget crises, two union negotiations, an LODD, and a department reorganization that nearly split the building in half. She's been, by every measure she knows how to apply, good at this job. She's been present when presence was what the job required. She's been willing when willingness was costly. She has been honest when honesty was uncomfortable.

And the city manager, who has worked with her for six of those eleven years, just told someone in a corridor that he doesn't need to talk to her anymore. Not because she failed. Because a system she helped configure gives him what she used to give him, faster, at seven in the morning, without requiring a meeting.

She sits with that for a long time.

The parking garage is quiet. A car passes somewhere below her. The fluorescent lights hum. She sits in the stillness of a person who just heard something they cannot unhear, and she doesn't try to think her way out of it. She just lets it be what it is.

She's not irreplaceable. She knew this, abstractly, from the first morning. But knowing it abstractly and hearing it said in a corridor are not the same thing. Hearing it said makes it the kind of knowledge that lives in your body rather than your head.

She thinks about the chief she spoke with — the furious one, twenty-three years in the same building, the scheduling tool he never evaluated. She thinks about how close the distance is between his situation and hers. She's been present in ways he has not, but the gap has been narrowing. The monthly briefing she stopped fighting for. The community advisory meeting a deputy has been attending in her place. The calls she's been meaning to make and has not made because the dashboard was there and the dashboard was sufficient.

The system did not create this problem. She can see that clearly, sitting in a parking garage, engine off. The system revealed it. The system made visible the distance she's been allowing to accumulate without naming it as distance, or as a problem.

She starts the car.

From the parking garage she calls the chief she knows three counties over.

He picks up on the second ring. He's been expecting a call — not from her specifically, but from someone. Version 5.0 has been doing the same things in his city that it has been doing in hers, and he's been sitting with a version of the same fear without the vocabulary to name it.

She doesn't name it for him immediately. She lets him talk. He talks about the unauthorized report distribution, the dashboard access, the analyst who cited a ChiefBot summary in a meeting without asking him first. His anger is the anger of someone who has been made redundant without being told, who has been moved out of the room quietly and professionally, in a way that makes complaint feel ungracious.

She recognizes it because she has been feeling a version of it for three weeks.

'How long has it been since you were in a station without a reason?' she asks.

The pause before he answers is long enough to be its own answer.

'A while,' he says.

'How long since a Council Member called you to ask you something they couldn't get from a report?'

Another pause.

'What are you saying?' he asks. Not hostile. Genuine.

'I'm saying I think the problem isn't the system,' she says. 'I think the system is just faster at the parts of our job we were already doing from a distance.'

He's quiet for a moment. Then: 'That's not a comfortable thing to say.

'No,' she agrees. 'It isn't.'

They talk for twenty minutes. She asks him about his department — not the metrics, the actual department. What is the thing his command staff is most worried about that doesn't appear in any report? He tells her. She listens. By the time they hang up she thinks he sounds slightly less like a man fighting for his seat and slightly more like a man trying to figure out what his job actually is.

She hopes she sounds the same.

She sits with the question for the rest of that day and all of the next.

What the portal gives the city manager is accurate information, delivered efficiently, on a schedule he controls. What it doesn't give him is what she heard in the corridor — not the words the city manager said, but the thing underneath them. The comfort of a leader who believes his department is running well and no longer feels he needs to verify that belief through conversation.

That comfort, Dana understands, is a risk. Not because the system is wrong. Because the system cannot tell him what it doesn't know, and there are things it doesn't know that matter. The hydraulic leak at Station 3, six weeks on a maintenance list. The firefighter driving twenty-six miles each way on a starting salary. The specific quality of discomfort in a station that has the highest attrition rate in the department and a captain who's never been told directly that what he is doing is *not* acceptable here.

None of that is in the portal. None of it will ever be in the portal. And as long as the city manager believes the portal is sufficient, none of it will reach him.

Unless she brings it.

Not in a briefing he has cancelled. Not in a report he can pull at seven a.m. In person. In the rooms where those things are visible. In the relationships that make those things legible to people who are not in those rooms.

The system cannot be the bridge between the city manager and the human texture of the department he funds. That's her job. It's always been her job. She's just been doing it badly enough, and just efficiently enough, that a more efficient system could make her look replaceable.

She opens her calendar.

She blocks two hours on Friday morning. Not for a briefing, not for a report, not for anything that can be generated automatically. She schedules herself to attend the Delmar and Fifth community advisory meeting in person, rather than sending a deputy as she has for the past four months.

Then she calls Council Member Park.

Not about the heat-related call volume. Not to add texture to information that has already. She calls because there's something she wants Park to know about the Delmar and Fifth district that's not in any report and will never be. She tells Park about the retired firefighter who volunteers at the station on weekends — grew up on Delmar, watched apparatus roll out of those doors as a child, was the first person from the neighborhood to join the department. She tells her about what it took to rebuild the department's relationship with the block association after the years it was allowed to deteriorate. She tells her what that relationship means to the station's capabilities in a targeted heat prevention program — the specific trust that makes a door-knock program viable rather than intrusive.

Park listens without interrupting. When Dana finishes, there's a pause.

'That's not in the summary,' Park says.

'No,' Dana says. 'It isn't.'

'It should be somewhere,' Park says. 'Who else knows this?'

‘The people who were there for it,’ Dana says. ‘And me.’

She drives to Station 3 on Saturday morning. Not an inspection. She brings nothing. She stays two hours.

She learns that the hydraulic leak has been moved up on the maintenance schedule because a firefighter filed a formal work order three days ago, which is what it took to get it noticed. She learns that the firefighter making the twenty-six-mile commute has been looking at apartments in the district and asked the battalion chief, two weeks ago, whether the department had any housing assistance resources — and the battalion chief did not know, and hadn’t followed up. She learns that the television called the hostage has been replaced — someone’s spouse dropped off a spare over the weekend, quietly, without announcement — and that the crew has been watching it together in the evenings in a way that has, fractionally, changed the temperature of the station.

Small things. None of them are in the data.

She thinks, driving back, about what it would cost to be here every week. Not the time — she has the time, she reclaimed it from the scheduling tool, she has been spending it on compliance reports she could have delegated. The discomfort, mostly. The willingness to sit in a dayroom where something is wrong and to not fix it immediately, to not turn it into an action item, to just be the chief who showed up and stayed long enough to learn something.

She’s been managing from behind a desk, using metrics and spreadsheets for two years. The metrics were right. They told her something was wrong. They could not tell her what the crew needed from her. That required being here.

The city manager calls on Monday morning.

His voice is different from the version she heard in the corridor. That version was efficient, satisfied, managing a system. This version is something closer to curious.

He'd been at the Delmar and Fifth community advisory meeting on Friday. He hadn't announced it — he does this occasionally when a community issue is on his radar and he wants to understand it without filtering. He had watched Dana in the room. The way she knew people by name. The way the block association chair had pulled her aside before the meeting started and shown her something on her phone, the two of them bent over it together in a corner, the chair pointing and Dana nodding. The way she had answered a question from a resident about apparatus response times with the kind of specific, quietly confident knowledge that has nothing to do with dashboards and everything to do with years of paying attention.

He had a question the portal couldn't answer. He wanted to know about the community's relationship with the station — not the metrics, the relationship. Whether it was real. Whether it would hold.

She told him. Then he asked about Station 3. She told him about the hydraulic leak, the commute, the housing assistance question the battalion chief hadn't followed up on. He was quiet for a moment after that.

'How did you know about the housing assistance question?'

'I was there Saturday,' she said.

Another pause. 'I pulled the Station 3 dashboard this morning,' he said. 'The numbers look like the numbers have looked for a year.'

'Yes,' she said. 'They do.'

The call ran thirty minutes. Near the end, without making it a formal request, he mentioned that the monthly briefing schedule had been harder to predict than he'd expected. He asked if she would be available to restart it.

'Second Thursday?' she offered.

'Second Thursday,' he agreed.

She set her phone down after the call and looked at her desk. The mug. The photograph she has been looking at since she sat in the parking garage three days ago.

She was not triumphant. She wasn't relieved in the way she'd expected. She was something quieter than that — the specific feeling of having found the right answer not by arguing for her seat at the table but by going to the places where she was irreplaceable and being present enough that someone noticed.

The dashboards could tell the city manager everything he needed to monitor the department. What they couldn't tell him was what the department meant to the people it served, or what it cost to serve them well, or what it looked like when a chief showed up on a Saturday morning with no agenda and stayed two hours.

That was hers. It had always been hers. The question — the only question that had ever mattered — was whether she would keep going.

Marcus Webb reads the same Version 5.0 upgrade notification on the same Thursday morning. He's been through this before — not this system, but this moment: the arrival of a capability that changes what the community, his firefighters, and elected officials need from him directly.

He reads the stakeholder transparency features section twice. He thinks about which of his Council Members will start pulling dashboards instead of calling him. He thinks about what those Council Members won't find in the dashboards.

He makes two notes in his calendar. Coffee with Council Member Osei-Bonsu, no agenda. Station 4 walkthrough, Tuesday, no announcement.

He doesn't feel threatened. He has been in enough buildings, in enough cities, to know that the right response to a system that can tell his stakeholders everything is not to compete with the system on its terms. It's to be, consistently and without fanfare, the one who can tell them what the system cannot.

He also holds a standing uncertainty meeting every other week with his command staff. Thirty minutes. No decisions. No action items. Just an honest inventory of the things the organization doesn't yet know how to do. The meetings have become, over time, one of the most trusted forums in the department — because they're the one place where not knowing is the expected condition rather than the embarrassing one.

Version 5.0 will be the agenda item at the next one.

He's curious what his team will say about it.

FOR THE READER

The question this chapter is asking is not whether AI will make you irrelevant. It's whether you will.

The chiefs who become invisible in the era of Version 5.0 are not the ones who fail to understand technology. They are the ones who have been measuring their value by the information they deliver, the reports they produce, the efficiency of their administrative presence — and who have been doing this for long enough that when a more efficient system arrives, their stakeholders barely notice the transition. The machine doesn't replace them. The distance they already created made them replaceable.

The chiefs who stay relevant are the ones who go where the machine can't. Not in a crisis. Consistently. As the ordinary baseline of how they lead.

1. Audit your last thirty days of contact with city leadership and elected officials. Not the reports you sent — the conversations you had. The calls you made that were not prompted by an agenda item. The briefings where you said something that wasn't in any document. The moments where your physical presence in a room changed what happened in that room. If those moments are scarce, Version 5.0 didn't create your visibility problem. It revealed it.

2. Identify three things about your department that aren't in any dashboard and will never be — the texture of a community relationship, the dynamic of a crew, the history behind a decision that looks obvious from the outside and is complicated from within. These are the things only you know because you have been present enough to learn them. They are your most durable source of professional value. Are you building them deliberately, or depleting them through distance?

3. Choose one thing you have been managing from a distance and go there this week. Not a formal visit. Not an inspection. Just go. Stay longer than is comfortable. Learn something the system couldn't have told you. Then think about what it would mean to do this consistently — not as an exceptional act, but as the ordinary baseline of how you lead.

The dashboards are accurate. The information is real.

What they cannot tell anyone is what you did on Saturday morning.

Chapter 11

The Future Chief

Scene

The conference is in a hotel ballroom in a city neither of them lives in.

Dana is presenting. She's been presenting at this conference for eleven years — the same ballroom, more or less, the same faces in the front rows, the same questions from the back. She knows this room the way she knows her apparatus bay. She knows where the sound dies in the far-left corner and which microphone cuts out after forty minutes and which moderator will let the Q&A run long if the conversation is good.

Before she begins, she stands at the front of the room while the audience settles as the A/V technician checks the clicker. She looks at the faces. Some she knows by name. Some she recognizes the way you recognize someone you've seen in professional contexts for years without ever having a real conversation. Some are new — younger faces, the next generation of company officers and battalion chiefs who are starting to fill the conference rooms that used to be exclusively occupied by people who looked like her and Marcus. She's been presenting in this room for eleven years and she's never stood at the front of it feeling quite like this. Not nervous. Something more specific than that. The sense that what she is about to say is going to land differently than what she's said in this room before

— and that differently is the right word, and she's not yet sure whether differently means better.

This year her presentation is different. Not the format — she's still the person at the front of the room, still the one with the slides and the case studies and the hard-won practitioner credibility that makes people lean forward. But the content has changed. She's talking about AI. Not abstractly, not theoretically — she's talking about what happened in her department over the past year. The report on the industrial corridor. The scheduling tool. The month the city manager stopped calling. The Station 4 memo. The community meeting at Delmar and Fifth. The two hours at Station 3 with bad coffee and a television called the hostage.

She's telling the truth about it — the vertigo and the resistance and the slow, imperfect learning. She's not presenting herself as someone who figured it out. She is presenting herself as someone who is still figuring it out, and who's learned enough to know that figuring it out is the job.

The room is more engaged than she has seen it in years.

Midway through the Q&A, she notices a woman in the third row who has been taking notes since before Dana began speaking. Not the half-attentive note taking of someone capturing key quotes — the focused, continuous writing of someone synthesizing everything in real time. She's young, or younger — early thirties, Dana estimates. She's wearing a battalion chief patch.

The woman raises her hand near the end of the session.

I want to ask about the community meeting, she says. When you scheduled it before you had the answer — what were you actually hoping to accomplish?

It's the best question Dana has been asked in eleven years of presenting.

She answers it. But she keeps thinking about it for the rest of the afternoon.

Battalion Chief Priya Nair drives back to her hotel after the session and opens her laptop.

She's been in the fire service for eleven years — the same amount of time Dana has been a chief. She made battalion chief at thirty-one, which was fast enough that some people noticed and a few people said things, and she's been carrying both the speed and the commentary ever since with the care of someone who's learned to take up exactly the right amount of space in a room.

She opens a document she's been keeping for eight months. It's titled, simply, Notes. It contains observations about AI tools she's used or evaluated, leadership decisions she's watched other chiefs make or not make, patterns she's noticed in how the fire service is absorbing — or failing to absorb — the technology arriving faster than most departments are prepared for.

She adds an entry.

Mercer gets it. Not all of it — nobody gets all of it yet. But she went to the community meeting before she had the answer. That's the thing. Most chiefs wait until they have the answer and then manage the community. She managed the uncertainty with them instead. That's different. That's what it will take.

Priya reads the entry back. She does this sometimes — reads what she's just written as if someone else wrote it, to check whether she believes it or only thinks she does. She believes this one. She's believed it since the Q&A, when Mercer answered the question about the community meeting and Priya recognized something she'd been trying to articulate for months: that the thing that separates the chiefs who will lead well through this transition from the ones who won't is not technical fluency. It's the willingness to be uncertain in public. To show up before you have the answer. To let the people affected by the decision be part of the process of making it. Mercer had done that instinctively, or maybe not instinctively — maybe at great cost, in the middle of something that was genuinely hard — and that was the distinction that mattered.

She closes the document. She opens the other one — the strategic framework draft she's been building for eight months without being asked. It's twenty-three pages. It outlines a human-AI partnership model for a mid-sized fire department: which decisions should remain exclusively human, which should be AI-informed, which should be AI-initiated with human oversight, and which — a short and carefully considered list — might eventually be entirely automated. Each category has governance criteria, accountability structures, and community transparency requirements. She's been testing the framework against real scenarios from her department's incident history, revising it each time it fails to produce a clear answer, building it toward something she believes the fire service will need and does not yet have.

She's not building it because someone told her to. She's building it because she can see what's coming and she can't stand arriving unprepared.

The fire service she's going to lead is already here. Most people just haven't seen it yet.

Story

The future of AI in the fire service is not a prediction. It's an inventory. The capabilities arriving in the next three to five years are not speculative. We're seeing extensions of what's already working, scaled and refined, and integrated. They'll change the operational baseline of what a fire department can do, and what a fire chief is expected to manage.

Understanding what's coming *is not optional* for the chief who intends to lead through it. It's the precondition for leading through it rather than being led through it by people who understand it better.

Computer-aided dispatch integration is already moving beyond simple call routing. AI systems are being integrated with CAD platforms to provide real-time recommendation engines — systems that analyze incoming call data, weather conditions, current apparatus positioning, historical response patterns, and available mutual aid resources simultaneously, and surface optimal dispatch recommendations in the seconds between a call coming in and apparatus being committed. The dispatcher and the incident commander still make the decision. The system is surfacing options that no human could generate at the same speed with the same completeness. Departments that have implemented these systems report measurable improvements in response time and resource utilization. Departments that have not are operating with less information in the same time window.

Predictive maintenance and apparatus health monitoring is moving from scheduled service intervals to continuous condition assessment. Embedded sensor arrays embedded in apparatus components transmit real-time data to AI systems that model failure probability across the vehicle's operational life. The system isn't waiting for a scheduled service interval to identify that a hydraulic system is trending toward failure — it identifies the trend weeks before the failure would occur, flags the apparatus for preventive service, and models the coverage impact of taking that unit out of rotation. For departments managing aging fleets with constrained capital budgets, this capability represents a fundamental shift in how apparatus lifecycle decisions are made.

Community risk prediction at the incident level is emerging as one of the most consequential applications in development. AI systems trained on incident history, building characteristics, occupancy patterns, and environmental factors are beginning to produce accurate predictions of which specific properties, in which specific conditions, represent elevated risk of a life-safety incident in the near term. This is not community risk reduction in the aggregate sense — it's the ability to identify the house on a specific street that, given its age, construction type, occupancy profile, and the current weather pattern, represents a meaningfully elevated probability of a fire in the next thirty days.

Consider what this looks like in practice: a department receives a system-generated list identifying a cluster of fourteen properties in a single neighborhood as elevated-risk — flagged not because of a recent incident, but because of a combination of structural age, deferred maintenance indicators, and seasonal occupancy patterns. The department deploys targeted smoke detector installation and home safety inspection resources to those properties over a six-week period. Several have detectors that are missing or non-functional. One has a heating system that inspectors flag for immediate service. The capability is not speculative. The Atlanta Fire Rescue Department has already implemented a version of this through the Firebird framework — a machine learning system developed in collaboration with researchers that computes fire risk scores across thousands of buildings, achieving true positive rates of up to 71 percent in predicting fires and identifying thousands of commercial properties not previously prioritized for inspection (Madaio et al., "Firebird: Predicting Fire Risk and Prioritizing Fire Inspections in Atlanta," arXiv:1602.09067 — available at https://arxiv.org/abs/1602.09067). The question is not whether this kind of precision is achievable. It is whether your department is prepared to use it.

Prevention resources deployed against that specificity of prediction are operating at a level of precision that general community risk analysis cannot approach.

Natural language incident documentation is already reducing the time cost of incident reporting and after-action documentation — and the timing matters. NERIS, the National Emergency Response Information System, requires API-based data submission and significantly more detailed, structured, and complete incident data than departments have ever been required to capture. Real-time submission is supported through RMS integrations but not mandated. Systems that can listen to radio traffic, CAD timestamps, and crew debriefs and generate structured incident reports — flagging gaps for human review rather than requiring human composition of the entire document — are moving from pilot programs into operational deployment, mapping output directly to NERIS fields. For departments where documentation burden falls disproportionately on company officers and overworks battalion chiefs, this capability returns meaningful time to the operational level of the organization while simultaneously improving the data quality that NERIS is designed to capture.

AI-assisted training scenario generation is making individualized, operationally relevant scenario development feasible at scales that were previously impractical. Systems that can generate high-fidelity scenario scripts, tabletop discussion guides, and skills evaluation frameworks from a department's own incident data — tailored to specific experience levels, specific skill gaps, specific incident types — are beginning to change what training officers can produce with the same number of hours.

What connects all these capabilities is not the technology. It's the leadership requirement they create.

Every one of these systems produces better outputs when it's given better inputs, better context, and better human judgment to govern its recommendations. Every one of them requires a chief officer who understands what the system is doing well enough to know when to trust it and when to override it. Every one of them creates decisions that didn't exist before

— decisions about data governance, about community transparency, about equity in how predictive systems are used, about the accountability structures that govern AI-assisted operational choices.

These are not technology decisions. They're leadership decisions. They require the same judgment, the same values orientation, the same relational intelligence, and the same organizational courage as every other leadership decision this book has described. The difference is that they are new — that the chief who faces them cannot draw on twenty years of accumulated organizational experience to navigate them, because twenty years ago they didn't exist.

This is Priya Nair's native terrain. She has been thinking about these decisions since before most of her peers knew they were coming. She's not smarter than Dana. She hasn't led longer than Marcus. She's simply grown up professionally in an environment where these questions were already present, and her relationship with them is fluency rather than adaptation.

The chief who's spent the past year doing what Dana has been doing — navigating the disruption, building the workflow, making the hard choices, going to Station 3 — is not behind Priya. They are on the same path, at a different point on it.

The chief who hasn't started is the one with something to worry about.

Priya is not impatient with the chiefs who are still figuring it out. She was a firefighter once, then a lieutenant, then a captain. She remembers what it felt like to be new to something that everyone around her seemed to already understand. She doesn't mistake her fluency for superiority.

But she's also honest about what she sees: a fire service that is moving toward a future it hasn't yet decided to prepare for, led in many departments by chiefs who are managing the present with frameworks built for the past. The gap between the capability that is arriving and the leadership capacity that is ready to receive it is real, and it's growing.

She's building the framework she wishes someone had handed her. She's documenting it, sharing it with the chiefs she trusts, presenting it at sessions that draw smaller rooms than Dana's but with people who are asking the same questions she was asking two years ago.

She's not waiting to be asked.

What she does with the recognition of the gap is the thing that separates her from the chiefs who see it and feel only its weight. She adds an entry to the Notes document. She sends the framework draft to a peer she's been corresponding with at a department in another state — a deputy chief who's been asking similar questions and building similar documents in parallel, without knowing the other person existed until a conference hallway conversation three months ago. She schedules a thirty-minute call for the following week to compare what they've built. This is how the future gets prepared for — not through institutional programs or mandated initiatives, but through the specific choices of specific people who cannot stand to arrive unprepared.

Dana drives back to her hotel that evening with the window down and the question still turning in her mind.

What were you actually hoping to accomplish?

She'd answered this question in the session — something about owning the uncertainty, about the community's right to be part of the decision before the decision was made, about what it means to lead with honesty rather than with answers. She meant all of it. But sitting with it now, alone on the highway, she thinks there was something underneath the answer she hadn't quite named.

What she'd been hoping to accomplish, she realizes, was to not be alone with it. The Station 4 decision was hers to make. But the weight of it — the history of Delmar and Fifth, the fragility of the trust that had been rebuilt, the specific cost of getting it wrong — that weight did not belong only to her. The community had been carrying it for decades. They deserved to carry the decision with her rather than having the decision delivered to them after the fact.

That's not a leadership framework. It's not a policy position. It's something closer to a conviction — the kind that forms slowly, through accumulated encounter, and is not fully articulable until someone asks exactly the right question at exactly the right moment.

Priya Nair asked exactly the right question.

That, Dana thinks, is the whole answer. Not the AI or the tools or the capability itself. The willingness to begin before you're ready, to learn in public, to go to the community meeting before you have the answer. The future is not arriving. It has arrived. The question is whether you're in it.

FOR THE READER

Three questions. Chapter 11 is not a warning. It's a window.

1. What AI capabilities are already present in your department's operational environment that you have not yet fully understood? Not the ones you are actively using — the ones that arrived, or are arriving, without your direct involvement. CAD integrations. Records management analytics. Dispatch optimization tools. Apparatus monitoring systems. Name them specifically. Then ask: do I understand what these systems are doing well enough to govern them well? Simply closing the door to their use or ignoring them doesn't

create a future-focused solution. If your answer – do I know enough to govern them – is no, this is step one. This is the gap to close first, before the next capability arrives.

2. Identify one emerging AI capability — from the list in this chapter or from your own research — that will materially affect how your department operates within the next three years. What decisions will it create that don't currently exist? Who in your department needs to understand it? What governance structure does it require? The chief who has thought through these questions before the capability arrives is in a fundamentally different position than the one who encounters it unprepared. It's the difference between proactive pre-planning and responding blind in the moment to the unusual facility.

3. Find your Priya. Not the person who agrees with you about technology — the person in your department or your professional network who's already living in the future you're still preparing for. The officer who's using AI tools in ways you haven't asked about. The peer at a different department building frameworks and agents you haven't seen. The person who would answer your questions about what's coming with specificity rather than speculation. Buy them coffee. Ask them what they see. Then listen without explaining why things are more complicated than they appear.

The future is not arriving.

It has arrived. The question is whether you are in it.

Conclusion

The Review

Scene

The report arrives on a Monday morning, ninety-three days after Municipal Leadership Optimization System, Version 4.2 was installed in the fire chief's office.

It is titled: 90-Day Performance Assessment — Chief Officer Function — Department of Fire Services.

The system has been collecting data since day one. Response time metrics. Administrative task completion rates. Budget variance. Training compliance. ISO documentation status. Grant submissions. Strategic plan progress against stated milestones. Community engagement activity, measured by scheduled events, attendance figures, and digital communication outputs. Personnel actions. Attrition rates by station.

The report is forty-one pages. It is formatted cleanly. It contains no grammatical errors. It is, by the system's own assessment, a comprehensive and accurate accounting of the chief officer function over the past ninety-three days.

Dana reads the executive summary first. The department is performing within acceptable parameters across most metrics. Response times are stable. Training compliance has improved six percent since the scheduling tool was implemented. The grant inventory has identified three new funding opportunities. The strategic plan is on track.

She reads through the operational section. Then the administrative section. Then the community engagement section, which notes two scheduled events: the Delmar and Fifth neighborhood meeting, and a follow-up session with the station advisory committee. Both are logged as completed. The community engagement metric has improved.

Then she reaches the personnel section.

> *Station 3 attrition rate: remains elevated at 2.3x department average. Supervisory relationship satisfaction: 31 percent below department average, unchanged from initial assessment. Recommended intervention: supervisory training program, follow-up survey. Status: no intervention implemented. Days since initial flag: 214.*

She reads the line three times.

214 days.

She has known for 214 days. She visited once — two hours, bad coffee, a television called the hostage. She learned three things the survey never captured. She has not been back. She has not talked to Reyes.

The system flagged it on day one. It has flagged it fourteen times since. It has no frustration about this. It has no awareness that anything has failed. It has simply noted, with the same precision it applies to everything, that the item remains open.

214 days.

She closes the report. She picks up her keys.

She does not call ahead.

Story: Dana

Station 3 is eleven minutes from headquarters on a Tuesday afternoon with no traffic.

Dana drives it in nine.

She drives the nine minutes without the radio. She has been composing this conversation in her head for 214 days and none of the versions she composed are available to her now. They were structures she built from a safe distance — careful openings, organized arguments, paths toward predetermined conclusions. Standing at the beginning of something real, she cannot find any of them. What she has instead is the knowledge that she should have come seven months ago and a clear-eyed understanding of what it cost the people at this station that she did not. She cannot walk into that watch office carrying either of those things as an excuse. She can only walk in carrying the decision to finally show up, and whatever follows from it.

She parks in the visitor lot rather than the chief's spot. She sits in the car for a moment — not long, just long enough to feel the weight of what she is about to do and decide she is going to do it anyway. Then she gets out.

The apparatus bay doors are open. Engine 3 is out on a call. The ladder is in. A firefighter she doesn't immediately recognize is wiping down the rear step with a cloth that has seen better days. He looks up when she comes in, and she sees the half-second recalibration — the recognition that this is the chief, the uncertainty about what that means for this particular afternoon.

Just stopping by, she says. Don't let me interrupt. He nods and goes to the station intercom: "Chief is here." She would have felt better without the station-wide announcement.

He nods and goes back to wiping down the step.

She finds Reyes in the watch office, standing at the scheduling board with a marker in his hand. He turns when she comes in and his face does the same half-second recalibration as the firefighter's, but with more history behind it.

Chief, he says.

Captain. She closes the door behind her. Do you have a few minutes?

He sets the marker down. Sure.

She does not sit down. Neither does he. They stand on opposite sides of a desk that has too much paper on it, and Dana thinks: this is it. This is the conversation she has composed in her head a hundred times and avoided for 214 days. The version in her head was always perfectly structured — the right opening, the clear statement of the problem, the path forward. Standing here now, she understands that the perfectly structured version was also a way of not having the conversation. It was comforting to take mental action — the sense of progress without the cost of going. There's no perfect structure for a conversation like this. There is only beginning.

I've been looking at the survey data from Station 3, she says. And the attrition numbers. And I've been getting reports for seven months telling me something isn't right here, and I want to talk to you about it directly rather than let it sit any longer.

Reyes's jaw tightens. She watches him process the choice he has — the defensive response that nine years of operating a certain way have made available to him, the deflection, the explanation that points outward rather than inward. She watches him almost choose it.

Then something shifts.

I know, he says. It's quiet. Two words. But they are the right two words, and they change everything about what the conversation can be.

When she asks him what he would do differently, he is quiet for a long time. Then he says: I would have asked more questions and given fewer answers. He says it looking at the scheduling board rather than at her. She understands that this is the most honest thing he has said to a chief officer in nine years at this station, and that it cost him something to say it, and that the right response is not to fill the space around it with affirmation or next steps or a performance plan. She lets it stand. After a moment he looks back at her, and she nods once, and that is enough.

She sits down. He sits down.

They talk for an hour and forty minutes.

She does not tell him he has been doing it wrong. She asks him where he learned to lead this way. He tells her — a department he came from fifteen years ago, a captain who ran the station through fear and called it discipline, a culture that rewarded the outcome without examining the method. He tells her he has known, for longer than he would like to say, that some of what he brought with him from that place does not belong here. He tells her three members who left in the past two years had told him directly — not in official channels, in conversations in this room — that they were leaving because of him. He tells her he thought about that more than he showed.

What follows is neither a performance improvement plan nor a disciplinary process. It is a conversation between two people, one of whom has been operating from a framework that was formed in another place and another time and has never been directly challenged, and one of whom has finally arrived to challenge it — not with punishment, but with honesty. Dana does not have all the answers. She tells him that. She tells him she has been figuring out her own relationship with how she leads over the past year, and that she does not have the standing to stand in judgment of a process that she has let run for seven months without showing up.

I should have come sooner, she says.

Yeah, Reyes says. But you came.

She leaves Station 3 at 4:47 p.m. She drives back to headquarters with the window down. She does not know yet how the story of this station ends. She does not know if Reyes will change, or how much, or how quickly, or whether the members who stayed will stay long enough to find out. She does not know if the conversation she just had was the beginning of something or simply the first step of a much longer road.

What she knows is that she went.

That, she thinks, is what she has been learning to do for the past year. Not to have the answers. To go.

Story: Marcus

Three hundred miles away, in a city Marcus Webb has lived in for fourteen months, Municipal Leadership Optimization System, Version 5.0 generates a similar report.

Marcus reads his in the morning, before anyone else arrives. He has been anticipating it — not with dread, but with the specific interest of someone who has been using a tool long enough to be curious about what it has noticed.

The report is accurate. His department is performing well across most metrics. There are two areas the system flags as requiring attention: a training compliance gap in one of the volunteer companies, and a budget variance in the apparatus maintenance line that is tracking slightly over projection.

And then: Community engagement — southern district: below department average and below prior-period baseline. Trend: declining over 90-day period.

Marcus sets the report down and thinks about the southern district. He had prioritized it heavily in his first year — the visits, the community organization meetings, the deliberate investment in neighborhoods where the department's relationship had been thin. The metrics had improved. He had moved on to other priorities.

He had stopped maintaining what he built.

He sits with that for a moment. This is the discipline he has been practicing for twenty-two years — not the avoidance of self-criticism but the willingness to receive accurate information about himself without flinching or rationalizing. The system found a pattern. The pattern is real. He allowed the southern district relationship to drift because the metrics had improved and he moved his attention to the next priority. That is a reasonable sequence of decisions. It is also the beginning of the distance he has spent a career warning other chiefs about. The fact that he caught it at ninety days rather than at nine months is the only thing that distinguishes him from the chief who does not catch it at all.

He picks up his phone and calls the director of a community center in the southern district — a woman he met eighteen months ago at a neighborhood association meeting, who had been cautious at first and warm by the end, who had called him twice since with questions about fire safety that were really questions about whether the department still cared about her neighborhood.

She picks up on the second ring.

Chief Webb, she says. I was wondering when I'd hear from you.

He winces. I owe you a visit, he says. Are you free Thursday?

There is a pause. Then: I'll make it work.

He puts the phone down. He makes a note in his calendar. He thinks about Dana, whom he has never met but whose department he has been tracking through a professional network they share — the reports that have filtered through, the decisions that have become case studies in the conversations he has been having with other chiefs about what this moment in the fire service actually requires.

He does not know her. But he recognizes her.

He opens the system interface and types a note into the community engagement file: Southern district — relationship maintenance lapsed. Course correction initiated. Check back in 30 days.

The system will flag it if he forgets. He does not intend to forget.

Story: Priya

Battalion Chief Priya Nair is in a meeting when her phone buzzes with a notification she has been setting up for three months — an automated alert from a grant monitoring system she configured herself, flagging a new federal funding opportunity for AI integration in fire service operations.

She silences the phone. She finishes the meeting. She reads the notification in the hallway afterward.

The grant is worth $340,000. The application window opens in six weeks.

She forwards it to the fire chief's office with a three-sentence summary and a note: This is worth pursuing. I've started the department profile section. Let me know if you want to discuss.

Then she goes back to work.

She is not waiting to be chief. She is practicing being one.

Closing

ChiefBot 9000 — Municipal Leadership Optimization System, Version 5.0 — does not know that Dana went to Station 3. It does not know what was said in the watch office, or that Reyes sat down, or that a conversation that should have happened seven months ago finally happened on a Tuesday afternoon because a line in a performance review moved someone to act.

It will notice, in ninety days, whether the attrition rate at Station 3 has changed. It will flag it if it has not. It will note the improvement if it has. It will update the metric and move to the next item in the queue.

It cannot know what it cost to get there.

It cannot know what it means.

It cannot go.

That is what the chief is for.

Go.

APPENDIX A

AI Glossary for Fire Service Leaders

This glossary defines terms used throughout this book and encountered regularly in conversations about AI and technology in the fire service. Definitions are written for practitioners, not technologists. Where a term has a precise technical meaning and a practical meaning, the practical meaning is given.

Terms are listed alphabetically. Where a term appears in italics within a definition, it has its own entry in this glossary.

Algorithm
A set of rules or instructions that a computer follows to complete a task or solve a problem. In the context of AI, algorithms are the underlying logic that determines how a system processes information and generates outputs. When a fire service AI tool recommends optimal apparatus positioning, an algorithm is performing the calculation.

Artificial Intelligence (AI)
A broad term for computer systems designed to perform tasks that typically require human-like reasoning — recognizing patterns, generating language, making predictions, or solving problems. AI is not a single technology; it is a category that includes many different approaches and tools. Modern

AI systems — particularly large language models — are broad-capability systems that can perform a wide range of tasks. What makes them useful in fire service leadership is not that they are narrowly specialized, but that they can be configured and prompted for specific applications with results that are immediately practical.

Augmentation
The use of AI to enhance human capability rather than replace it. An AI tool that helps a chief officer draft a grant narrative faster is augmenting the chief's work. An AI tool that drafts the narrative and submits it without human review is automating it. This distinction — augmentation versus automation — matters throughout this book. See also: Automation.

Automation
The use of technology to perform a task without human involvement. Automation is appropriate for rules-based, repeatable tasks where human judgment adds little value — generating a compliance report from existing data, for example. It is not appropriate for tasks that require judgment, context, or accountability. The goal of AI in fire service leadership is augmentation first, automation where it is genuinely appropriate. See also: Augmentation.

CAD Integration
The connection of computer-aided dispatch systems with AI tools that can analyze incoming call data, current apparatus positioning, historical response patterns, and other variables to surface optimized dispatch recommendations in real time. CAD-integrated AI does not replace the dispatcher or incident commander — it provides better information in less time.

Community Risk Prediction
The use of AI to identify specific properties or geographic areas at elevated risk of a fire or life-safety incident based on building characteristics, occupancy patterns, incident history, and environmental factors. Distinguished

from general community risk assessment by its specificity — rather than identifying high-risk neighborhoods, community risk prediction can identify high-risk individual properties within a defined time window.

Context Document

A brief, structured description of a department's key characteristics — size, staffing model, response area, accreditation status, apparatus complement, union status, policy formatting conventions — used as input when working with AI tools. Providing a context document at the beginning of an AI conversation significantly improves the relevance and usability of the output. See Chapter 9.

Data Governance

The policies, procedures, and accountability structures that determine how data is collected, stored, accessed, used, and protected. As AI tools generate and use more departmental data, data governance becomes a leadership responsibility — determining who can access what information, how AI-generated outputs are reviewed before distribution, and how the department protects sensitive personnel and community data.

Decision Support System

An AI tool designed to provide information and recommendations that support human decision-making rather than replace it. Most fire service AI applications are decision support systems — they surface patterns, flag anomalies, generate options, and provide analysis. The human remains accountable for the decision. This is the appropriate model for the vast majority of fire service AI applications.

Generative AI

AI systems that produce new content — text, images, audio, code — in response to prompts. Large language models like ChatGPT and Claude are generative AI. In fire service applications, generative AI is used for drafting reports, policies, grant narratives, training scenarios, and correspondence. The output requires human review before use.

Hallucination

A term used to describe AI output that is confidently stated but factually incorrect. Large language models can generate plausible-sounding text that contains errors, invented citations, or false information. This is a known limitation of current AI technology, with research suggesting hallucination rates in high-stakes domains can run between 8 and 15 percent in current systems. The risk is not uniform: it is highest in tasks requiring precise factual accuracy — citations, regulatory references, specific statistics — and lower in tasks like first-draft narrative or structural outlining where a human editor reviews the output. Providing a strong context document and reviewing AI-generated content against source material significantly reduces hallucination risk. It is the primary reason that AI-generated content — especially in official documents, grant applications, or policy — must be reviewed by a knowledgeable human before use.

Human-AI Partnership Model

A framework for determining which decisions in an organization should remain exclusively human, which should be AI-informed, which should be AI-initiated with human oversight, and which might be automated entirely. Developing a human-AI partnership model is an emerging leadership responsibility for fire chiefs. See Chapter 11.

Large Language Model (LLM)

The technology underlying AI tools like ChatGPT, Claude, Gemini, and others — a list that is illustrative, not exhaustive, as new models from additional developers appear regularly. LLMs are trained on enormous amounts of text and can generate coherent, contextually appropriate written responses to prompts. They are the category of AI most immediately relevant to fire service administrative and documentation work. Their primary limitation is the hallucination problem — they can generate confident errors. Capability and reliability are improving rapidly; a tool evaluated in one year may perform substantially differently in the next. See also: Hallucination, Generative AI.

Lifecycle Management

In the context of apparatus, the use of AI to model the full cost trajectory of a vehicle — maintenance history, manufacturer specifications, parts availability trends, peer department replacement benchmarks — and generate replacement schedules that optimize both operational readiness and budget predictability. AI-assisted lifecycle management moves apparatus replacement decisions from reactive (when it breaks) to strategic (when the data says it should). See Chapter 4.

Machine Learning

A type of AI that improves through experience — the system processes data, identifies patterns, and adjusts its outputs over time without being explicitly reprogrammed. Machine learning underlies many fire service AI applications, including predictive response analytics, apparatus health monitoring, and community risk prediction. The more data the system processes, the more accurate its outputs tend to become.

Natural Language Processing (NLP)

The branch of AI concerned with understanding and generating human language. NLP is what allows AI tools to read a written incident report, extract structured information from it, respond to a prompt written in plain English, or generate a policy document. Most AI tools that fire chiefs interact with directly rely on NLP.

Predictive Analytics

The use of historical data and statistical modeling to forecast future outcomes. In fire service applications, predictive analytics is used for response time optimization, apparatus maintenance scheduling, community risk prediction, budget forecasting, and staffing projections. The output of predictive analytics is a probability, not a certainty — it tells you what is likely, not what will happen.

Prompt

The input given to an AI tool — a question, instruction, or description that tells the tool what to produce. The quality of a prompt directly determines the quality of the output. Vague prompts produce vague results. Specific prompts with clear context, constraints, and output requirements produce usable results. Prompt literacy — the ability to write effective prompts — is the most transferable skill in working with AI tools. See Chapter 9.

Prompt Engineering

The practice of designing and refining prompts to produce better outputs from AI tools. In fire service applications, prompt engineering is less about technical sophistication and more about specificity — providing the tool with accurate department context, clear instructions, and explicit output requirements. See also: Prompt, Context Document.

Records Management System (RMS)

Software used by fire departments to store, organize, and retrieve operational records — incident reports, personnel files, training records, apparatus maintenance logs, and related data. AI tools that integrate with an RMS can analyze patterns across large datasets that would be impractical to review manually, enabling the kind of data-driven insights described throughout this book.

Risk Modeling

The use of data analysis to quantify and prioritize risk across a department's response area. AI-assisted risk modeling draws from incident history, building characteristics, demographic data, infrastructure condition, and other variables to produce ranked assessments of where life-safety risk is highest. Risk modeling informs prevention program deployment, inspection prioritization, and facilities planning.

Training Data

The dataset used to train an AI system — the information the system learns from before it is deployed. The quality, scope, and representativeness of training data directly affects the accuracy and reliability of the system's outputs. In fire service AI applications, understanding what data a system was trained on — and what its limitations therefore are — is part of responsible use.

Workflow Automation
The use of AI to perform recurring, rules-based tasks automatically — generating scheduled reports, flagging compliance lapses, distributing notifications, updating maintenance logs. Workflow automation is the category of AI most immediately accessible to most fire chiefs and most likely to return significant time with modest implementation effort. See Chapter 9.

This glossary covers terms as they are used in this book and in current fire service AI applications. The field is evolving rapidly. Definitions that are accurate today may require revision as the technology and its applications develop. When in doubt, verify current usage against the most recent resources from IAFC, FEMA, and NFPA. Terms and processes related to incident documentation should be verified against the most current NERIS (National Emergency Response Information System) guidance, as fire service data standards and terminology continue to evolve.

APPENDIX B

AI Tools Reference for Fire Service Administrators

This appendix provides an annotated reference to AI tools and tool categories relevant to fire service administration, organized by function. It is not an exhaustive product directory — it is a practical starting point for chiefs who want to understand what is available and where to begin.

Tools listed here were current and available at the time of writing. The AI tools landscape changes rapidly. Verify current availability, pricing, and features before making adoption decisions. Where specific products are named, inclusion does not constitute endorsement — the goal is to illustrate the category with concrete examples.

A standing disclaimer applies to this entire appendix: AI capabilities, pricing, and availability evolve continuously. Always verify current information directly with vendors and against your department's data governance and procurement requirements before implementation.

Category 1: General-Purpose Large Language Models

These tools handle a wide range of text-based tasks — drafting, summarizing, analyzing, answering questions, generating scenarios. They are the most accessible entry point for most fire chiefs and the tools most immediately useful for administrative work.

ChatGPT (OpenAI)
One of the most widely used large language models. Available in free and paid tiers. Useful for drafting reports, policies, grant narratives, training scenarios, correspondence, and analysis. The paid version (ChatGPT Plus) provides access to more capable models and additional features including web browsing and document analysis.

Primary use: *Report drafting, policy development, grant narrative writing, training scenario generation, correspondence, research synthesis.*

Claude (Anthropic)
A large language model with strong performance on long-document tasks and nuanced writing. Available in free and paid tiers. Particularly useful for working with lengthy documents — policy manuals, strategic plans, after-action reports — and for tasks requiring careful, structured prose. Maintains context well across long conversations.

Primary use: *Policy review and drafting, strategic plan development, long-form report writing, complex analysis, document summarization.*

Gemini (Google)
Google's large language model, integrated with Google Workspace tools including Docs, Sheets, and Gmail. Useful for departments already operating within the Google ecosystem. Strong performance on research tasks and document-based work.

Primary use: *Document drafting within Google Workspace, research synthesis, data analysis in Google Sheets, correspondence.*

Microsoft Copilot

Microsoft's AI assistant, integrated with Microsoft 365 applications including Word, Excel, Outlook, and Teams. For departments operating on Microsoft platforms, Copilot can assist with document creation, spreadsheet analysis, email drafting, and meeting summaries without leaving familiar applications.

Primary use: *Document drafting in Word, data analysis in Excel, email assistance in Outlook, meeting summaries in Teams.*

Category 2: Scheduling and Staffing Optimization

These tools process staffing constraints, leave requests, certification requirements, and minimum staffing standards to generate optimized shift schedules. They reduce manual scheduling time significantly and catch conflicts that manual processes miss.

Aladtec

A workforce management platform designed specifically for public safety agencies, including fire departments. Handles scheduling, time and attendance, availability tracking, and compliance reporting. Widely used in the fire service with integrations for common payroll systems. Note: Aladtec is now part of TCP Software — search under either name when researching the product.

Primary use: *Shift scheduling, leave management, overtime tracking, minimum staffing compliance, certification expiration alerts.*

ScheduleFlex / Scheduling Software Platforms

A category of scheduling tools — including ScheduleFlex, TCP Humanity (formerly Humanity, now also under the TCP Software umbrella), and others

— that use optimization algorithms to generate schedules from staffing rules and constraints. Not all are fire-service-specific, but many can be configured for the unique requirements of twenty-four-hour shift operations.

Primary use: *Schedule generation, conflict identification, coverage gap analysis, overtime minimization.*

Category 3: Records Management and Data Analytics

These tools connect to department records management systems and provide analytical capabilities that surface patterns, trends, and insights from incident, training, and operational data.

ESO Fire RMS
A records management system with built-in analytics capabilities for fire departments. Tracks incident data, personnel information, training records, and apparatus maintenance. Analytics features allow pattern analysis across incident types, response times, and personnel performance. ESO Fire RMS is actively building NERIS readiness into its platform — a relevant consideration for departments aligning with current NERIS requirements.

Primary use: *Incident data analysis, response time tracking, training compliance monitoring, personnel performance data.*

ImageTrend
A data management and analytics platform used widely in EMS and fire service. Provides reporting tools, dashboard capabilities, and data integration with CAD systems. Useful for chiefs wanting to move from annual reporting to continuous performance monitoring.

Primary use: *Incident reporting, data analytics, CAD integration, performance dashboards, compliance reporting.*

Power BI (Microsoft) / Tableau
General-purpose data visualization and analytics platforms that can connect to fire service data sources and create custom dashboards and reports. Require more configuration than fire-service-specific tools but offer greater flexibility. Power BI is accessible at low cost for departments already on Microsoft platforms.

Primary use: *Custom performance dashboards, data visualization, trend analysis, budget modeling, community risk mapping.*

Category 4: Training Design and Scenario Generation

These tools assist training officers in developing scenario-based exercises, curriculum outlines, and evaluation frameworks. They reduce the time required to develop individualized training content.

General-Purpose LLMs (see Category 1)
ChatGPT, Claude, and similar tools can generate scenario scripts, tabletop discussion guides, skills evaluation checklists, and curriculum outlines when given specific incident types, learning objectives, and participant experience levels as input. The output requires review and refinement by a subject matter expert but reduces development time significantly.

Primary use: *Scenario script generation, tabletop exercise development, curriculum outline drafting, evaluation checklist creation.*

Lexipol
A policy management and training platform for public safety agencies. Provides policy development tools, training modules tied to policy content, and compliance tracking. AI-assisted features are being integrated across the platform. Widely used in fire departments for policy management.

Primary use: *Policy development, training compliance, policy-tied training content, regulatory update monitoring.*

Category 5: Grant Research and Application Development

These tools help departments identify grant opportunities and develop competitive applications. They reduce the research burden and improve narrative quality.

Grants.gov / FEMA Grant Finder
Federal grant portals that list available funding opportunities. Not AI-powered, but the starting point for any federal grant search. FEMA's AFG, SAFER, and FP&S programs are listed here. Pairing these portals with a general-purpose LLM for application drafting is a practical and accessible approach for most departments.

Primary use: *Federal grant opportunity identification, application requirement review, deadline tracking.*

Instrumentl
A grant research and management platform that uses AI to match organizations with grant opportunities based on their profile and priorities. Provides eligibility screening, deadline tracking, and application management. Particularly useful for departments that pursue multiple grant sources and need to manage the pipeline systematically. Note: Instrumentl is primarily designed for nonprofit organizations. Fire departments that operate as government entities rather than 501(c)(3) nonprofits should verify eligibility for grants in its database before subscribing — many opportunities listed may require nonprofit status.

Primary use: *Grant opportunity identification, eligibility matching, application tracking, funder research.*

General-Purpose LLMs for Application Drafting
ChatGPT, Claude, and similar tools can draft grant application narrative sections when provided with accurate department information and the grant's scoring criteria. The chief provides the context document, the grant requirements, and the department's specific need — the tool produces a working draft that the chief refines and verifies against the current Notice of Funding Opportunity before submission. Always verify AI-drafted content against current program requirements.

Primary use: *AFG narrative drafting, financial need documentation, cost-effectiveness justification, project description development.*

Category 6: Community Risk and Prevention Analytics

These tools help departments analyze community risk data and deploy prevention resources more precisely.

ESRI ArcGIS
The leading geographic information system platform, widely used in fire service community risk reduction. AI-assisted features are being integrated for predictive risk mapping, resource deployment optimization, and community vulnerability analysis. Many departments already have ArcGIS access through municipal GIS departments.

Primary use: *Community risk mapping, response area analysis, prevention resource deployment, demographic overlay, coverage gap identification.*

RiskCall / Community Risk Reduction Platforms
A category of specialized tools designed to help fire departments conduct community risk assessments, manage prevention programs, and track outcomes. Tools in this category — including RiskCall and similar platforms

— connect incident data with occupancy information to identify high-risk properties and populations.

Primary use: *Home safety assessment tracking, smoke detector program management, high-risk occupancy identification, prevention program outcome measurement.*

Getting Started: A Practical Sequence

For chiefs who are new to AI tool integration, the following sequence minimizes complexity while maximizing early return:

Month 1: Begin with a general-purpose LLM (ChatGPT or Claude, free tier). Build your department context document. Use it to draft one recurring administrative document — a monthly report, a policy section, or a grant inquiry letter. Evaluate the output. Refine your prompts.

Month 2: Identify your highest-volume scheduling or compliance tracking burden. Evaluate one purpose-built tool in that category. Most offer free trials. Configure it with your department's parameters and run it in parallel with your existing process for one scheduling cycle before fully transitioning.

Month 3: Identify one data analytics gap — a question about your department's performance that you cannot currently answer with available reports. Work with your RMS vendor or a general-purpose analytics tool to build one dashboard that addresses it.

Months 4-6: Add integrations based on what is returning the most value and consuming the most time. The goal at the end of six months is a functional workflow covering scheduling, compliance tracking, report drafting, and at least one analytics application — not perfection, but operational infrastructure.

Tools and pricing verified at time of writing. The AI tools landscape evolves rapidly. Verify current availability, features, and pricing directly with vendors before making procurement decisions. Consult your department's data governance policies and legal counsel regarding data privacy requirements before connecting any AI tool to departmental records systems.

APPENDIX C

Personal Adaptability Self-Assessment

This assessment is a reflective tool, not a clinical instrument. It is not designed to produce a score that categorizes you or predicts your performance. It is designed to surface patterns in your relationship with change that are worth examining honestly — patterns that may be invisible to you precisely because they are familiar.

Answer each question as you actually are, not as you would like to be. The assessment is for you. Its value is entirely dependent on your honesty with yourself.

Use the following scale for each question:

1—Rarely or never 2—Sometimes 3—Often 4—Almost always

Section 1: Career History and Context

These questions establish context. There are no right or wrong answers. They are designed to help you see your own professional history clearly.

1. How many different organizations have you worked in during your career?

(This is a number, not a rating. Note it here:)

Consider this number as context for interpreting your responses throughout the rest of the assessment. Research on leadership adaptability consistently shows that broader organizational experience correlates with higher adaptive capacity \u2014 not because more departments is inherently better, but because each new environment requires recalibrating assumptions that the previous environment made invisible. A chief who has spent a full career in one department has deep institutional knowledge and may have exceptional adaptive capacity; they may also have had fewer forced opportunities to examine their foundational assumptions. A chief who has worked in six departments may have strong adaptive habits and a weaker grasp of what sustained presence in a community actually builds. Neither is a verdict. Erika Ayers Badan, in Nobody Cares About Your Career (2024), makes a related argument from a career development perspective: that being new, repeatedly and deliberately, builds a kind of professional agility that long tenure in a single organization \u2014 for all its depth and institutional mastery \u2014 can quietly erode. The veteran who knows every system and every unspoken rule brings irreplaceable value; they may also have stopped noticing what a fresh set of eyes would immediately question. This number is a lens, not a score. Use it to read your own patterns honestly.

2. When you encounter a new organizational policy, procedure, or tool, your first instinct is curiosity rather than resistance.

1—Rarely or never *2—Sometimes* *3—Often* *4—Almost always*

3. You can identify at least three specific things you have changed about how you lead in the past five years — not things you have added, but things you have changed because you found a better way.

1—Rarely or never 2—Sometimes 3—Often 4—Almost always

4. When something in your department has always been done a certain way, you regularly ask whether it should continue to be done that way.

1—Rarely or never 2—Sometimes 3—Often 4—Almost always

5. You actively seek out information about how departments other than your own handle challenges similar to yours.

1—Rarely or never 2—Sometimes 3—Often 4—Almost always

Section 2: Response to Change

These questions examine how you actually respond when change arrives — not how you believe you should respond.

6. When a new technology, policy, or approach is introduced in your department, your default response is to evaluate it before forming an opinion about it.

1—Rarely or never 2—Sometimes 3—Often 4—Almost always

7. When something you have done successfully for years is challenged or replaced, you are able to examine the challenge honestly rather than defending the existing approach reflexively.

1—Rarely or never *2—Sometimes* *3—Often* *4—Almost always*

8. You have recently changed your mind about something significant — a leadership approach, a policy position, an operational assumption — based on new information or evidence.

1—Rarely or never *2—Sometimes* *3—Often* *4—Almost always*

9. When you feel resistant to something new, you examine the source of your resistance before acting on it.

1—Rarely or never *2—Sometimes* *3—Often* *4—Almost always*

10. You distinguish between resistance that is based on principled objection and resistance that is based on discomfort with the unfamiliar.

1—Rarely or never *2—Sometimes* *3—Often* *4—Almost always*

Section 3: Learning and Development

These questions examine the practices that either build or erode adaptive capacity over time.

11. You read or engage with professional content outside your immediate area of expertise or operational focus.

1—Rarely or never *2—Sometimes* *3—Often* *4—Almost always*

12. In the past year, you have deliberately placed yourself in a learning situation where you did not have established expertise — a new training program, an unfamiliar conference track, a peer department with a different operational model.

1—Rarely or never *2—Sometimes* *3—Often* *4—Almost always*

13. You are comfortable not knowing the answer to a question in front of your team and saying so directly.

1—Rarely or never *2—Sometimes* *3—Often* *4—Almost always*

14. You seek feedback from people who are likely to challenge your thinking, not only from people who are likely to affirm it.

1—Rarely or never *2—Sometimes* *3—Often* *4—Almost always*

15. When you make a mistake in how you led something, you examine what the mistake reveals about your assumptions rather than focusing primarily on the external factors that contributed to it.

1—Rarely or never *2—Sometimes* *3—Often* *4—Almost always*

Section 4: Technology and AI Specifically

These questions examine your current relationship with AI and technology change in the fire service.

16. You have personally used an AI tool for a professional task in the past six months.

1—Rarely or never *2—Sometimes* *3—Often* *4—Almost always*

17. You understand what the AI tools currently operating in your department are doing well enough to govern them — to know when to trust their output and when to question it.

1—Rarely or never *2—Sometimes* *3—Often* *4—Almost always*

18. When AI tools produce results that differ from your instinct or experience, your first response is to examine the difference rather than dismiss the tool's output.

1—Rarely or never *2—Sometimes* *3—Often* *4—Almost always*

19. You have a clear sense of which decisions in your department should remain exclusively human, which should be AI-informed, and which could be AI-initiated with human oversight.

1—Rarely or never *2—Sometimes* *3—Often* *4—Almost always*

20. You are actively preparing for AI capabilities that will arrive in your department in the next three years, not only managing the ones that have already arrived.

1—Rarely or never *2—Sometimes* *3—Often* *4—Almost always*

Section 5: Organizational Leadership

These questions examine how you lead your organization through change, not only how you adapt personally.

21. Your department has a culture in which members feel safe raising concerns about current practices without fear of professional consequences.

1—Rarely or never *2—Sometimes* *3—Often* *4—Almost always*

22. You visibly model learning — your team sees you encountering something new, trying it, getting it wrong, and course-correcting.

1—Rarely or never *2—Sometimes* *3—Often* *4—Almost always*

23. When your department resists a change you believe is necessary, you engage with the resistance directly rather than either forcing the change or abandoning it.

1—Rarely or never *2—Sometimes* *3—Often* *4—Almost always*

24. Your department's training program reflects current incident data rather than historical tradition. It changes when the data changes.

1—Rarely or never *2—Sometimes* *3—Often* *4—Almost always*

25. Members of your department at all levels would describe the organization as one that learns — that takes in new information, adjusts its practices, and improves over time.

1—Rarely or never *2—Sometimes* *3—Often* *4—Almost always*

Scoring and Reflection

Add your scores for questions 2 through 25. (Question 1 is contextual — it does not have a score.) Your total will fall between 23 and 92.

23 to 45 — Your relationship with change has significant rigidity. This is not a verdict on your competence or your commitment to the fire service. It is information about where you are starting. The questions that produced your lowest scores are the most useful starting points for deliberate development.

46 to 68 — You have a working relationship with change that functions adequately but has identifiable gaps. Look at the questions where you scored 1 or 2. Those are the areas where your adaptive capacity is most constrained — and most likely to limit you when the next significant disruption arrives.

69 to 92 — Your relationship with change is generally healthy. The questions that still produced 1s and 2s are worth your attention. No one scores 92. The goal is not a perfect score — it is honest self-knowledge and deliberate development.

Regardless of your score, the most important question in this assessment is the last one:

What is one thing you have been resisting that you cannot explain beyond the fact that it is different?

Write it down. Sit with it. That is where your development begins.

This assessment is also available as a downloadable resource at TrainTeachLead.com, where it can be used with department leadership teams as a group reflection tool.

APPENDIX D

Recommended Reading and Resources

The books and resources in this appendix are organized into four categories that correspond to the book's central themes: fire service leadership and culture, technology adoption and organizational change, continuous learning and professional development, and AI and the future of work. Each entry includes a brief annotation explaining its relevance to the arguments in this book.

This is not a comprehensive bibliography. It is a curated starting point — the resources most likely to extend the thinking this book has begun and challenge the assumptions this book has not been able to fully examine.

Fire Service Leadership and Culture

The Five Dysfunctions of a Team — Patrick Lencioni (2002)
The leadership fable that most directly influenced the form of this book. Lencioni uses a fictional executive team to illustrate real and measurable organizational dysfunction. Relevant here not only for its argument but for its demonstration that story is a legitimate and effective vehicle for serious leadership content. If you found this book's approach effective, Lencioni's work is the clearest antecedent.

Leadership and Self-Deception — The Arbinger Institute (2000)
An examination of how leaders undermine their own effectiveness through self-justifying thinking — the gap between how we see ourselves and how we actually behave. Directly relevant to Dana's arc in this book and to the broader argument about what chiefs avoid and why. Challenging in the best sense.

Turn the Ship Around! — L. David Marquet (2013)
A true story of how Navy Captain David Marquet transformed the USS Santa Fe from the worst-performing submarine in the fleet to the best by replacing the traditional leader-follower model with a leader-leader model. USA Today named it one of the 12 best business books of all time; Fortune called it the best how-to manual anywhere for managers on delegating, training, and driving flawless execution. Directly relevant to the culture chapter and to the broader argument about what fire service leadership should produce in the people it touches. The leader-leader model translates directly to fire service organizations where decision-making at the company officer level determines operational outcomes.

Extreme Ownership — Jocko Willink and Leif Babin (2015)
A case for radical leadership accountability drawn from military experience. The core argument — that the leader owns everything in their environment, including failures they did not directly cause — maps directly onto the Dana character arc and the book's central argument about avoidance. Read critically; some of its operational prescriptions translate to the fire service better than others.

Multipliers: How the Best Leaders Make Everyone Smarter — Liz Wiseman (2010)
An examination of how some leaders amplify the intelligence and capability of those around them while others diminish it. Highly relevant to the mentorship and belonging sections of Chapter 7 and to the broader argument about what fire service leadership actually produces in the people it touches.

Technology Adoption and Organizational Change

The Innovator's Dilemma — Clayton Christensen (1997)
The foundational work on how successful organizations fail to adapt to disruptive technology — not because they make bad decisions, but because they make good decisions optimized for a world that is changing faster than they recognize. Required reading for any fire chief trying to understand why smart, competent departments resist change that is clearly beneficial.

Diffusion of Innovations — Everett Rogers (1962, updated 2003)
The research foundation for understanding how new ideas and technologies spread through organizations and communities. Explains why some members adopt new approaches early, why most wait to see what happens, and why some never adopt at all. Provides a framework for leading technology change that is grounded in how change actually moves through human systems.

The Culture Code: The Secrets of Highly Successful Groups — Daniel Coyle (2018)
An examination of what makes certain groups — sports teams, military units, creative organizations — dramatically more effective than others. Coyle identifies three core skills: building safety, sharing vulnerability, and establishing purpose. Directly relevant to the culture chapter and to the Station 3 arc throughout the book. Practical, research-grounded, and immediately applicable to fire service company and chief officer culture building.

Thinking in Systems — Donella Meadows (2008)
A foundational text on systems thinking — the ability to see organizations as interconnected systems rather than collections of independent parts. Essential for chiefs who want to understand why culture change is hard, why well-intentioned interventions sometimes produce unexpected results, and why the same problem keeps recurring despite repeated attempts to fix it.

Continuous Learning and Professional Development

Mindset: The New Psychology of Success — Carol Dweck (2006)
The research behind the growth mindset — the finding that people who believe their abilities can be developed through dedication and hard work achieve more than those who believe their abilities are fixed. Directly relevant to the adaptability argument in Chapter 10 and to the broader case that leadership capacity is developed through practice, not innate.

Think Again: The Power of Knowing What You Don't Know — Adam Grant (2021)
An argument for the importance of rethinking — updating beliefs, assumptions, and practices in the face of new evidence. Grant argues that the most effective leaders and organizations build cultures of productive disagreement and intellectual humility. Directly relevant to the adaptability argument in Chapter 10 and to the broader case that the most dangerous thing in a fire department is the belief that what has worked in the past will continue to work in the future. One of the most practically useful books on this list.

Atomic Habits — James Clear (2018)
A practical framework for building the habits that produce long-term change. Relevant to the workflow-building argument in Chapter 9 and to the broader point that adaptability is a practice, not a trait. The most immediately actionable book on this list for chiefs who want to translate the arguments in this book into daily behavior.

Nobody Cares About Your Career: Why Failure Is Good, the Great Ones Play Hurt, and Other Hard Truths — Erika Ayers Badan (2024)
A candid, experience-grounded guide to how work actually works, written by the former CEO of Barstool Sports who built her career across multiple organizations before leading one. Badan's own trajectory — from Microsoft and AOL through a series of pivots to an unlikely leadership role — informs a core argument that runs throughout the book: that being new,

repeatedly and deliberately, builds a kind of professional agility that long tenure in a single organization, for all its deep institutional knowledge, can quietly erode. The veteran who knows every system, every workaround, and every unspoken rule of one organization brings irreplaceable depth; they may also have stopped noticing what a fresh set of eyes would immediately question. Badan does not diminish institutional expertise. She argues that awareness of its limits is itself a form of expertise. Directly relevant to the adaptability arguments in Chapter 10 and to Question 1 of the self-assessment in Appendix C — and a useful counterweight to any chief who has been in one building long enough to stop asking why things are done the way they are done.

The Infinite Game — Simon Sinek (2019)
An argument that the most effective leaders play an infinite game — building organizations designed to outlast any individual and focused on sustained purpose rather than short-term wins. Sinek distinguishes between finite players who play to win and infinite players who play to keep the game going. Directly relevant to the staying relevant argument in Chapter 10 and to the question of what fire service leadership is actually for when the administrative work has been assumed by AI.

AI and the Future of Work

Co-Intelligence: Living and Working with AI — Ethan Mollick (2024)
One of the most practically useful and accessible books on working with AI written for a general professional audience. Mollick is a business professor who has studied AI integration extensively and writes with both rigor and clarity. Directly relevant to the prompt literacy and workflow arguments in Chapter 9. Recommended as the first book a chief reads after finishing this one.

The Coming Wave — Mustafa Suleyman (2023)
A serious examination of where AI technology is heading and the governance challenges it creates — written by one of the founders of DeepMind. More technical than most books on this list but important for chiefs who want to understand the trajectory of the capabilities described in Chapter 11. Not a fire service book, but the leadership and governance arguments translate directly.

Power and Prediction: The Disruptive Economics of Artificial Intelligence — Ajay Agrawal, Joshua Gans, and Avi Goldfarb (2022)
A rigorous examination of how AI redistributes power within organizations and across industries — specifically, the shift from AI as a prediction tool to AI as a decision-making force. Written by three economists who study AI and its organizational implications. More practically grounded than most books in this category. Directly relevant to the governance arguments in Chapters 8 and 11 and to the question of which decisions should remain with the chief and which can be AI-assisted.

The Alignment Problem — Brian Christian (2020)
A journalist's examination of the challenges involved in building AI systems that reliably do what humans intend. More accessible than Russell's book and equally relevant to the ethical and governance arguments in this book. Particularly useful for chiefs developing frameworks for which decisions should remain human and which can be AI-assisted.

Fire Service Research and Data

IAFC Officer Development Handbook — International Association of Fire Chiefs (Current edition)
The fire service's primary resource for chief officer competency development. Relevant as a baseline against which to measure the new competencies this book argues are emerging — particularly around technology literacy, data-driven decision making, and community engagement.

NFPA 1750 — National Fire Protection Association (Current editions) The standards governing career and volunteer fire department operations, including response time objectives, staffing requirements, and apparatus deployment. Understanding these standards is essential context for evaluating AI-assisted decisions about resource allocation, facilities, and staffing — particularly the Station 4 scenario in Chapter 8.

Fire Engineering Magazine and Firefighter Nation — ongoing resources for fire service practitioners. Both publish regular content on technology, leadership, and operational topics relevant to the arguments in this book. The author is a contributing writer to Fire Engineering.

TrainTeachLead.com — the author's professional development platform for fire service leaders. Includes a dedicated AI for the Fire Service resource hub (trainteachlead.com/ai-for-the-fire-service/), downloadable tools including the Appendix C self-assessment, the Fire Service AI Explorer, the Leadership Conversation Simulator, and ongoing content on AI, leadership, and organizational development in the fire service.

ChiefBot9000.com — the companion website to this book. Includes the Personal Adaptability Self-Assessment, additional tools for chief officers navigating the territory this book describes, and resources for departments ready to move from reading to doing.

Publication dates reflect original editions. Many of these works have been updated or have newer editions available. Verify current editions before purchase. All resources listed were available at time of writing.

APPENDIX E

Research Basis and Source Notes

This book is a leadership fable. Its characters, departments, cities, and scenarios are fictional — composite constructions designed to illuminate real dynamics rather than document specific events. The story is the delivery mechanism. The arguments inside it are grounded in genuine research, practitioner experience, and published data.

This appendix identifies the research foundation underlying the book's principal claims, distinguishes between fictional illustrations and factual arguments, and directs readers who want to go deeper to primary sources. It is not a formal academic bibliography — it is a transparent accounting of where the book's evidence comes from.

On the Fictional Elements

Chief Dana Mercer, Chief Marcus Webb, Battalion Chief Priya Nair, Captain Reyes, Firefighter Okafor, ChiefBot 9000, and all departments, cities, council members, and scenarios in this book are invented. Any resemblance to actual persons, departments, or events is coincidental.

Illustrative statistics within the narrative — Marcus Webb's prevention program producing a measurable decline in residential fires, the scheduling tool returning twelve hours per month — are constructed illustrations, not citations. They are designed to be plausible and directionally accurate but should not be treated as documented findings. The community risk prediction scenario in Chapter 11 was revised to reference a real implementation: the Atlanta Fire Rescue Firebird framework (Madaio et al., arXiv:1602.09067), which achieved true positive rates of up to 71 percent in predicting fires. Where specific statistics are cited in the text, sources are identified below.

Research Foundation by Chapter

The following notes identify the research basis for the book's principal factual claims, organized by chapter.

Introduction and General Framework

1. The claim that AI can replicate significant portions of administrative, analytical, and strategic fire chief functions is supported by a growing body of practitioner literature and vendor documentation. The author's assessment of current AI capabilities in the fire service draws from direct experience with available tools, IAFC Technology Council publications and Technology Summit International proceedings, and product documentation from major fire service software vendors including ESO, ImageTrend, and Lexipol.

Chapter 1: Administrative Burden

2. Chapter 1 references McKinsey research finding that managers across industries spend roughly one full day per week on administrative tasks, paired with the author's practitioner observation that fire chiefs carry a heavier version of this burden due to ISO documentation cycles, NERIS reporting, accreditation requirements, compliance tracking, and grant management obligations. The McKinsey figure is directionally consistent with fire service leadership development literature. No single published study specific to fire chief administrative time burden was identified at a confidence level sufficient for direct citation; the argument rests on the cross-industry research base combined with the author's professional experience and practitioner consensus.

Chapter 3: Training and Psychological Safety

3. The discussion of psychological safety as a measurable organizational condition draws from Amy Edmondson's foundational research, particularly her 1999 paper 'Psychological Safety and Learning Behavior in Work Teams' published in Administrative Science Quarterly, and her subsequent book The Fearless Organization (2018). The application of psychological safety research to high-reliability organizations (HROs) including fire service operations is supported by subsequent literature in the HRO field.

4. The reference to high-reliability organization research draws from Karl Weick and Kathleen Sutcliffe's Managing the Unexpected (2001, updated 2015) and related HRO literature. The fire service is increasingly recognized in this literature as a relevant context for HRO principles, though the direct application of HRO research to AI integration in fire departments is an emerging rather than established body of work.

Chapter 4: Strategic Planning and Grants

5. The description of FEMA's Assistance to Firefighters Grant program, Staffing for Adequate Fire and Emergency Response program, and Fire Prevention and Safety grant program reflects program parameters current at the time of writing. Grant program priorities, funding levels, and eligibility criteria change annually. All AFG-related claims should be verified against the current Notice of Funding Opportunity published by FEMA prior to any application.

Chapter 5: Firefighter Suicide

6. The claim that firefighters die by suicide at rates exceeding line-of-duty deaths from fire and trauma combined is supported by multiple sources. Primary sources cited in the text: the Ruderman Family Foundation white paper on mental health and first responders; and the Firefighter Behavioral Health Alliance (FBHA), which documented 112 confirmed firefighter suicides in 2024 and estimates that confirmed cases represent approximately 60 percent of actual cases. Supporting research includes: Tiesman HM, Konda S, Hendricks S, et al. (2015) 'Suicide in U.S. Workplaces, 2003-2010' American Journal of Preventive Medicine; Jahnke SA, Poston WSC, Haddock CK, et al. (2016) 'Firefighting and mental health' and related publications in the occupational health literature; and ongoing surveillance data published by the Firefighter Cancer Support Network and the National Fallen Firefighters Foundation. The author acknowledges that methodological variation across studies produces somewhat different rate estimates; the directional finding — that suicide represents a significant and often undercounted source of firefighter mortality — is consistently supported.

Chapter 8: Equity and Community Risk

7. The argument that fire departments sometimes underserve economically vulnerable communities and majority-minority communities in prevention and inspection resource deployment is supported by a body of research on fire risk disparities. Key sources include work by the Home Fire Campaign, published analyses of NFIRS data by USFA and NFPA examining fire risk by socioeconomic and demographic factors, and community risk reduction literature examining resource allocation patterns. The fictional Station 4 scenario is constructed to illustrate a real and documented pattern rather than to document a specific case.

Chapter 10: Adaptability Research

8. The argument that adaptability is a developed capacity rather than a fixed trait draws from Carol Dweck's mindset research (see Appendix D), Adam Grant's work on rethinking and intellectual humility (Think Again, 2021), and general cognitive science literature on expertise and transfer learning. The specific application to fire service leadership — that single-organization tenure can erode adaptive capacity — is the author's inference from this research base applied to fire service career patterns, not a finding from direct empirical study of fire chiefs.

Chapter 11: Emerging AI Capabilities

9. The five emerging AI capabilities described in Chapter 11 — CAD integration, predictive maintenance, community risk prediction, natural language documentation, and AI-assisted scenario generation — are based on the author's assessment of current product development trajectories, vendor roadmaps, and pilot program reports available at the time of writing. These are not speculative technologies; they represent extensions of capabilities already operational in some departments. Community risk prediction specifically is grounded in the Atlanta Fire Rescue Firebird framework: Madaio M, Chen S-T, Haimson OL, et al. (2016) 'Firebird: Predicting Fire Risk and Prioritizing Fire Inspections in Atlanta,' arXiv:1602.09067 (https://arxiv.org/abs/1602.09067). Natural language documentation is developing to meet the API-based submission requirements of NERIS, the National Emergency Response Information System, and its more detailed data standards; AI documentation tools are being designed to support them. Availability, cost, and maturity vary significantly across vendors and implementations.

The Author's Published Research

The arguments in this book are supported by the author's professional experience, published practitioner work, and ongoing engagement with fire service research communities. Resources directly relevant to the book's content: trainteachlead.com/ai-for-the-fire-service — AI resources, tools, and practical guidance for fire service leaders, developed alongside this book. trainteachlead.com/research-library — the author's published research archive, including peer-reviewed work in fire service occupational health, cancer presumption law, and industrial firefighter workforce analysis. A 57-jurisdiction statutory dataset is archived on Zenodo (doi:10.5281/zenodo.18521370). These publications establish research credentials in the fire service but are not directly cited in this book's narrative. trainteachlead.com/published-work — practitioner articles and conference presentations in Fire Engineering, Firefighter Nation, and related publications. ChiefBot9000.com — the companion website to this book, with additional tools and resources for chief officers applying the book's arguments.

A Note on AI Assistance

This book was written with AI assistance, the same way it recommends using it. The tools used my data and story to help refine drafts and develop structure. The judgment, the arguments, and the responsibility for every claim within it are mine.

Readers seeking additional source material, the author's published datasets, or supplementary resources are encouraged to visit TrainTeachLead.com and the author's Zenodo repository. The author welcomes correspondence from researchers and practitioners working on related questions.

About the Author

Alex Zielinski is a Fire Chief at a major industrial facility in Indiana with more than nineteen years of experience across municipal, industrial, and volunteer organizations, including company and chief officer roles focused on training, performance, and emergency preparedness.

Alex serves on the IAFC Industrial Fire & Safety Section and is a student in the National Fire Academy's Executive Fire Officer Program. He has contributed to Fire Engineering, Firefighter Nation, JEMS, and published peer-reviewed research in the Journal of Occupational and Environmental Medicine.

Alex presents on leadership, technology, and high-performance firefighting at conferences including FDIC, Fire-Rescue International, and the Alberta Fire Chiefs Association Conference. He created TrainTeachLead.com as a professional development resource for fire service leaders.

Outside the firehouse, Alex is a Cincinnati Reds loyalist and lifelong baseball fan. He's also a former music teacher and drummer/percussionist.

www.ingramcontent.com/pod-product-compliance
Lightning Source LLC
LaVergne TN
LVHW050815061126
851857LV00012B/810
9798996613601